EAI/Springer Innovations in Communication and Computing

Series editor
Imrich Chlamtac, European Alliance for Innovation, Gent, Belgium

Editor's Note

The impact of information technologies is creating a new world yet not fully understood. The extent and speed of economic, life style and social changes already perceived in everyday life is hard to estimate without understanding the technological driving forces behind it. This series presents contributed volumes featuring the latest research and development in the various information engineering technologies that play a key role in this process.

The range of topics, focusing primarily on communications and computing engineering include, but are not limited to, wireless networks; mobile communication; design and learning; gaming; interaction; e-health and pervasive healthcare; energy management; smart grids; internet of things; cognitive radio networks; computation; cloud computing; ubiquitous connectivity, and in mode general smart living, smart cities, Internet of Things and more. The series publishes a combination of expanded papers selected from hosted and sponsored European Alliance for Innovation (EAI) conferences that present cutting edge, global research as well as provide new perspectives on traditional related engineering fields. This content, complemented with open calls for contribution of book titles and individual chapters, together maintain Springer's and EAI's high standards of academic excellence. The audience for the books consists of researchers, industry professionals, advanced level students as well as practitioners in related fields of activity include information and communication specialists, security experts, economists, urban planners, doctors, and in general representatives in all those walks of life affected ad contributing to the information revolution.

Indexing: This series is indexed in Scopus and zbMATH

About EAI

EAI is a grassroots member organization initiated through cooperation between businesses, public, private and government organizations to address the global challenges of Europe's future competitiveness and link the European Research community with its counterparts around the globe. EAI reaches out to hundreds of thousands of individual subscribers on all continents and collaborates with an institutional member base including Fortune 500 companies, government organizations, and educational institutions, provide a free research and innovation platform.

Through its open free membership model EAI promotes a new research and innovation culture based on collaboration, connectivity and recognition of excellence by community.

More information about this series at http://www.springer.com/series/15427

Yujie Li • Huimin Lu
Editors

3rd EAI International Conference on Robotic Sensor Networks

ROSENET 2019

Springer

EAI
RESEARCH MEETS INNOVATION

Editors
Yujie Li
Fukuoka University
Fukuoka, Japan

Huimin Lu
School of Engineering, Kyushu Institute of Technology
Fukuoka, Japan

ISSN 2522-8595 ISSN 2522-8609 (electronic)
EAI/Springer Innovations in Communication and Computing
ISBN 978-3-030-46031-0 ISBN 978-3-030-46032-7 (eBook)
https://doi.org/10.1007/978-3-030-46032-7

This Springer imprint is published by the registered company Springer Nature Switzerland AG.
The registered company address is: Gewerbestrasse 11, 6330 Cham, Switzerland

Preface

We are delighted to introduce the proceedings of the 2019 European Alliance for Innovation (EAI) International Conference on Robotic Sensor Networks (ROSENET2019). The theme of ROSENET 2019 was "Cognitive Internet of Things for Smart Society." This proceedings highlights selected papers presented at the 3rd EAI International Conference on Robotic Sensor Networks held in Kitakyushu, Japan.

Today, the integration of artificial intelligence and Internet of Things has become a topic of growing interest for both researchers and developers from academic fields and industries worldwide. Artificial intelligence is poised to become the main approach pursued in the next-generation IoTs research.

The rapidly growing number of artificial intelligence algorithms and big data devices has significantly extended the number of potential applications for IoT technologies. However, this also poses new challenges for the artificial intelligence community. The aim of this conference is to provide a platform for young researchers to share the latest scientific achievements in this field, which are discussed in these proceedings.

The technical program of ROSENET 2019 consisted of 9 full papers from 24 submissions. Aside from the high-quality technical paper presentations, the technical program also featured three keynote speeches. The three keynote speeches were Prof. Hyoungseop Kim, Prof. Masuhiro Nitta, and Prof. Yuya Nishida from Kyushu Institute of Technology, Japan.

Coordination with the steering chair, Imrich Chlamtac, was essential for the success of the conference. We sincerely appreciate his constant support and guidance. It was also a great pleasure to work with such an excellent organizing committee team for their hard work in organizing and supporting the conference. In particular, the Technical Program Committee, led by our Program Chair, Dr. Shenglin Mu, Dr. Jože Guna, and Dr. Shota Nakashima, who have completed the peer-review process of technical papers made a high-quality technical program. We are also grateful to the Conference Manager, Lukas Skolek, for his support and all the authors who submitted their papers to the ROSENET 2019 conference and special sessions.

We strongly believe that ROSENET conferences provide a good forum for all researchers, developers, and practitioners to discuss all science and technology aspects that are relevant to Robotics and Cognitive Internet of Things. We also expect that the future ROSENET conferences will be as successful and stimulating as indicated by the contributions presented in this volume.

Fukuoka, Japan Yujie Li
Fukuoka, Japan Huimin Lu

Conference Organization

Panels Chair
Guangwei Gao — Nanjing University of Posts and Telecommunications, China

Demos Chair
Hao Gao — Nanjing University of Posts and Telecommunications, China

Tutorials Chairs
Csaba Beleznai — Austrian Institute of Technology, Austria
Ting Wang — Nanjing Tech University, China

Technical Program Committee

Dong Wang	Dalian University of Technology, China
Weihua Ou	Guizhou Normal University, China
Xing Xu	University of Electronic Science and Technology of China, China
Haiyong Zheng	Ocean University of China, China
Rushi Lan	Guilin University of Electronic Technology, China
Zhibin Yu	Ocean University of China, China
Wenpeng Lu	Qilu University of Technology, China
Yinqiang Zheng	National Institute of Informatics, Japan
Jinjia Zhou	Hosei University, Japan

Contents

Reinforcement Learning-Based Cell Intelligent Multimode Frequency Reuse Method

Zhiling Tang, Yuan Wang, Rushi Lan, Xiaonan Luo, and Simin Li

1 Introduction

Inter-cell interference exists in the upstream and downlink. Orthogonal Frequency Division Multiple Access (OFDMA) is used in current and next-generation wireless communication systems. Because signals in the cell are orthogonal to each other, intra-cells interference is avoided [1]. However, users at the edge of a cell may use the same time-frequency resource block to generate co-channel interference to the neighboring cell, which is called inter-cell interference (ICI) [2]. As shown in Fig. 1, user 1 in the cell and user 2 in the neighboring cell use the same frequency carrier to transmit an uplink signal to the respective base station. The signal transmitted by user 2 would interfere base station A, and the signal from user 1 to the base station A would interfere base station B. These may seriously affect the signal-to-interference-plus-noise ratio (SINR) which leads to a decline in communication quality. In Fig. 2, there is the inter-cell downlink interference model. The base station A sends a signal to user 1 and also receives the same frequency signal sent by the base station B. But for user 1, it is also an ICI.

It can be seen from Figs. 1 and 2 that user 1 and user 2 are located at the edge of the respective cells and vulnerable to be covered by the inter-cell interference than users near the base station. This difference leads to the lower throughput of the entire cell network. Moreover, there is inhomogeneity in the quality of service (QoS) that the central users may be superior to the edge users. Therefore, improving the QoS of the edge users by mitigating ICI is a future important task.

Frequency reuse (FR) is considered to be an effective method to solve ICI [3–5]. The frequency band is divided into several parts, which are respectively allocated

Z. Tang · Y. Wang · R. Lan (✉) · X. Luo · S. Li
Guangxi Key Laboratory of Wireless Broadband Communication and Signal Processing, Guilin University of Electronic Technology, Guilin, Guangxi, China

Y. Li, H. Lu (eds.), *3rd EAI International Conference on Robotic Sensor Networks*, EAI/Springer Innovations in Communication and Computing,
https://doi.org/10.1007/978-3-030-46032-7_1

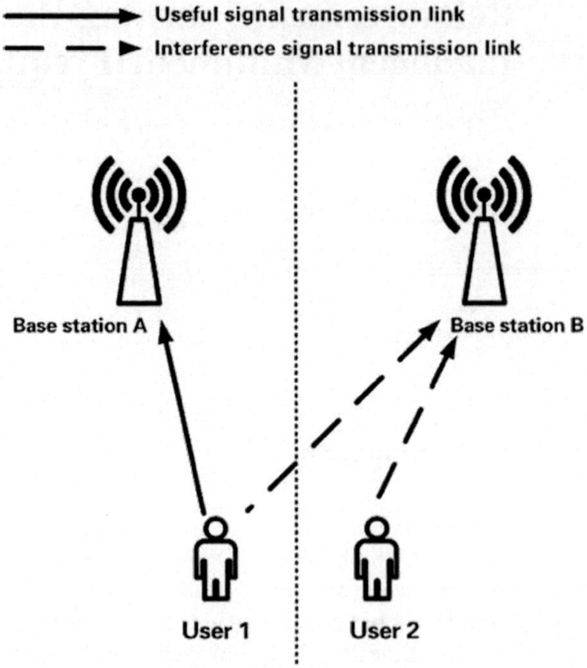

Fig. 1 Uplink interference of small interval interference model

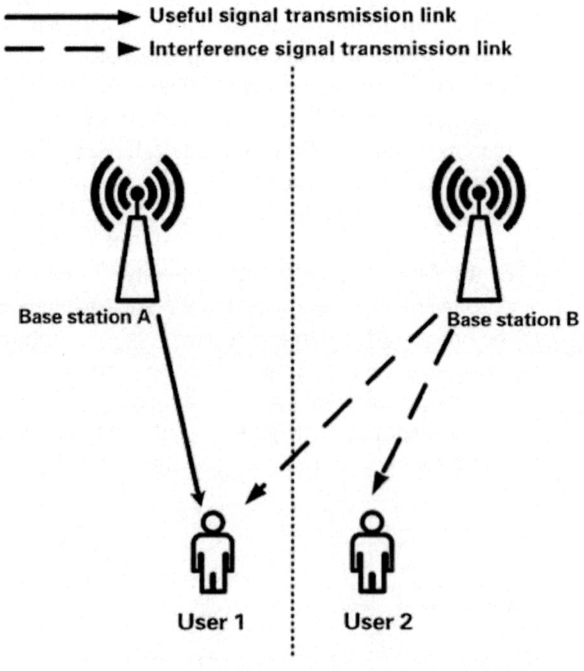

Fig. 2 Downlink interference of small interval interference model

to the central users and edge users and ensure the edge users of the neighboring cells using different frequency bands [6, 7]. According to the way of frequency allocation, FR can be divided into static FR and dynamic FR. The static FR has low complexity and low network signaling overhead and is easy to implement in engineering. Dynamic frequency reuse method can dynamically modify frequency reuse according to interference intensity, network load, network coverage, and other conditions. But its disadvantage is that the network overhead is large and the algorithm is complex.

In many cases, the reduction of interference is at the cost of the reduction of spectral efficiency or energy efficiency. An advanced energy efficiency measure for FR is proposed [8]. An adaptive partial frequency reuse scheme, which divides a cell into two different virtual regions, is proposed [9]. With more deployed femtocells to reuse available resources as much as possible, less resources can be allocated to the portion of the macro-cell [10]. By classifying users by the principle of avoiding major interference sources, different users use different reuse factors to improve the throughput of the whole system [11]. To avoid interference and crosstalk, an adaptive frequency reuse scheme for femtocell networks is used [12]. Exhaustive search and greedy descent methods are used to find the subcarrier and power allocations in each cell and then are iteratively repeated among cells until a predefined convergence criterion is satisfied [13]. The multilevel soft FR scheme gives 2N power density limit levels, achieving a better interference pattern and further improving the cell edge and the overall data rate [14]. In multiuser scenarios, the service time of the network is minimized through the rational allocation of resources [15].

In this chapter, an intelligent multimode FR scheme based on reinforcement learning is proposed. Compared to traditional frequency reuse methods, this FR method is more flexible for different wireless channel state. The reinforcement learning-based scheme can enable optimal FR modes under different channel conditions. Its goal is to improve the SINR of received signals and throughput.

2 Intelligent Multimode FR

FR is a concept proposed by Bell Labs in 1947, which is the cornerstone of cellular mobile communication. FR factor represents the number of frequencies in a reuse cluster. The larger the FR factor, the larger the reuse distance. Moreover, the analysis shows that the larger the FR factor, the lower the spectrum efficiency and the smaller the interference [16, 17]. Because the inter-cell co-channel interference decreases with the increase of distance, the interference of the central users is smaller than that of the edge users. The strategy of dividing central users and edge users is critical to the performance of cellular networks.

In fact, the location of a user and the number of central users and edge users are time-varying and random in the mobile cellular network. If the number of edge users increases, and the FR mode with the FR factor of 3 is allocated, the requirements of all users cannot be satisfied. Since a single frequency reuse cannot meet the requirement, we propose an intelligent multimode frequency reuse method based on reinforcement learning. Through reinforcement learning mechanism, one of the four frequency reuse modes is selected to optimize network throughput, energy efficiency, and spectrum utilization.

Firstly, the definitions of some main parameters that determine the mode of frequency reuse are given. It is assumed that the state of the cellular network is synchronized. The base station is in the transmitting state, and the user is in the receiving state. ICI will be generated if users in the neighboring cell use the same spectrum resource. The SINR is

$$\mathrm{SINR} = \frac{P}{(I_{\mathrm{own}} + I_{\mathrm{other}}) + N} \tag{1}$$

where P is the useful signal power received by the user, I_{own} is interference inside the cell, I_{other} is interference between adjacent cells, and N is noise. P is defined as follows:

$$P = P_{bs} + G - \mathrm{P}_{\mathrm{loss}} - P_{\mathrm{shadow}} \tag{2}$$

where P_{bs} is the transmit power, G is the gain of the antenna, P_{loss} is the path loss, and P_{shadow} is the decline of shadow. Because OFDMA for downlink enables orthogonality between symbols, the intra-cell interference is avoided. This leads to the simplified SINR:

$$\mathrm{SINR} = \frac{P}{I_{\mathrm{other}} + N} \tag{3}$$

1. Throughput K_{th}

   Throughput is related to the modulation method [18]. It is assumed that Quadrature Phase Shift Keying (QPSK) modulation is adopted, and the coding rate is 1/2. Then, throughput is defined as

$$K_{th} = \frac{n \times R_b}{T_f} \tag{4}$$

   where n is the number of subcarriers included in each resource block, R_b is the bit rate of each subcarrier, and T_f is the frame length.

2. Coverage K_{cover}

   If a SINR value is taken as a threshold γ, the percentage of the total users who receive signals greater than or equal to γ is defined as coverage:

$$K_{\mathrm{cover}} = \frac{\text{Number of users } (\gamma \geq 1)}{\text{Total users}} \tag{5}$$

3. Spectrum efficiency K_{eff}

   The spectral efficiency is as follows:

$$K_{eff} = \frac{K_{th}}{B} \tag{6}$$

   where B is the channel bandwidth.

4. Spectrum utilization factor U_{spe}

   In the FR scheme, the FR factor is defined as K_{fr}. The effective utilization of the spectrum resources reflects the size of the reuse factor, which is defined as

$$U_{\text{spe}} = \frac{1}{S * P_{\max}} \sum_{i \in S}^{n} P_i \tag{7}$$

   where S is the total number of sub-bands, $P_{\max}$ is the maximum transmitted power of all sub-bands, and P_i is the transmitting power that belongs to the sub-band i. The soft FR factor is reciprocal to the spectrum utilization factor as follows:

$$K_{fr} = \frac{1}{U_{\text{spe}}} \tag{8}$$

If the bandwidth of the edge region is 1/3 of the total bandwidth and the transmitting power is $P_{\max}$, and the bandwidth of the central region is 2/3 of the total bandwidth and the transmitting power is alpha $\alpha P_{\max}(0 \le \alpha \le 1)$, the following equation can be obtained:

$$U_{\text{spe}} = \frac{1}{S * P_{\max}} \left(\frac{S}{3} P_{\max} + \frac{2S}{3} \alpha P_{\max} \right) = \frac{1 + 2\alpha}{3} \tag{9}$$

It can be seen from Eqs. (8) and (9) that both the spectrum utilization rate and the soft FR factor can be expressed by the transmission power ratio α. The larger the α, the smaller the FR factor and the higher the spectrum utilization rate. According to Shannon's theorem, the capacity of edge users is $C_e = (B/3)\log_2(1 + SINR_{\text{edge user}})$, and that of center users is $C_c = B\log_2(1 + SINR_{\text{center user}})$. The FR factor of the central user is 1, which can maximize the spectrum utilization efficiency, and Eq. (7) shows that the throughput is maximized. Due to the serious interference to the edge users, only by increasing the FR factor can SINR be enhanced, so the coverage area of the cell has been greatly improved. However, according to Eq. (9), the spectrum resources used are 1/3 of the total bandwidth of the system. When the throughput of the central user is equal to that of the edge user, then

$$B\log_2 \left(1 + SINR_{\text{edge user}}\right) = \frac{B}{3}\log_2 \left(1 + SINR_{\text{center user}}\right) \tag{10}$$

Therefore,

$$SINR_{\text{edge user}} = (1 + SINR_{\text{center user}})^3 - 1 \quad (11)$$

Equation (11) shows that the SINR of edge users must be larger than that of central users under the same capacity. In the scheme of partial FR, the division of center radius has a great influence on the system capacity of the cell.

According to the above analysis, four FR modes are proposed:

(a) The FR factor is equal to 1. The spectrum utilization rate is the highest, but the edge users are seriously disturbed and the coverage rate is the lowest.
(b) Soft FR mode. The frequency band consists of high power band and low power band. The high power band is twice the low power band.
(c) Soft FR mode. The frequency band consists of high power band and low power band. The high power band is four times the low power band.
(d) The FR factor is equal to 1/3. Although the spectrum utilization rate is low and the interference of the same frequency is small, the coverage rate becomes larger.

Reinforcement learning neural network (RLNN) is used to select one of the four frequency reuse modes which is most suitable for the current cellular network. For the dynamic cellular environment, RLNN considers the SINR on its input, which is called "Detection Neural Network" (DeNN), as shown in Fig. 3. DeNN is a combination of multiple deep neural networks, and all deep neural networks are trained in parallel by the same training data set, and finally, multi-objective performance prediction values are generated. Each NN in DeNN has a feedforward structure trained by Levenberg-Marquardt backpropagation algorithm [19]. It consists of three fully connected unbiased layers: two hidden layers consist of 5 and 50 neurons (weighted parameters of each neural network are 349), two layers use log-sigmoid

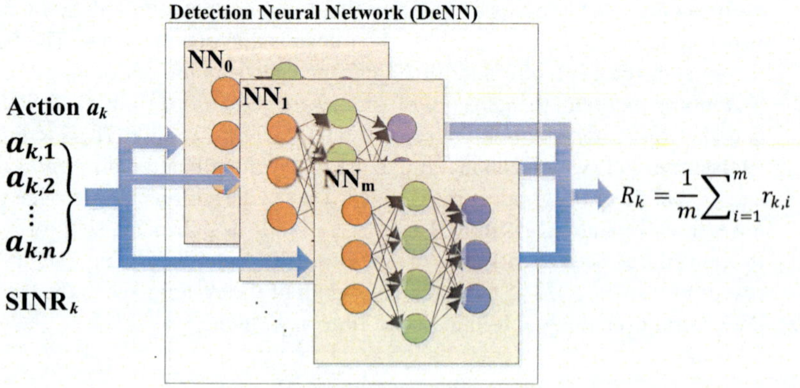

Fig. 3 Deep NNs used by reinforcement learning detection receive the same multidimensional input

transfer function, and output layer uses standard linear transfer function in each neuron. It uses the mean squared error (MSE) as the performance function under two different early stop training conditions: the minimum error gradient is equal to 10^{-12}, and the maximum validation checks is equal to 20. During the training process, the training data sets are randomly divided into 70% for training, 15% for testing, and 15% for validation, and all data are scaled to (0, 1). DeNN contains 20 NNs trained using the same training data set. The predictive output of DeNN is used to classify input actions according to the performance threshold of FR mode selector agent. The "action rejection probability" control randomly selects an action from a good or bad set, which in turn is used to select FR mode.

3 Results

In system simulation, some of the parameters are adaptable. The range described in Table 1 follows the sequence of operations described in FR control algorithm. Multidimensional actions consisting of five parameters a = (MT;$\overline{c}$; $\overline{M}$; m; c) are used in dynamic cell environment.

The simulation conditions are shown in Table 2.

We compared intelligent multimode FR with other FR methods and discussed their advantages and disadvantages. In addition, the resource scheduling algorithm

Table 1 DeNN adaptable parameters

Parameter	Variable	Value range
Modulation type	$\overline{MT}$	[QPSK,QAM]
Coding rate	$\overline{c}$	[1/3,1/2,2/3,3/4,4/5]
Modulation order	$\overline{M}$	[4,8,16,32]
Bandwidth	$\overline{B}$	[2.5–10] MHz
FR factor	$\overline{FR}$	[1–3]

Table 2 The simulation conditions

Parameter	Value
Cell topology	Macro cell
Number of cells	19
Station distance	1000 m
Number of cellular users	5000
Total number of resource blocks	50
BS transmit power	46 dBm
Fading type	Large-scale fading, fast fading
Path model	128.1 + 37.6lg(d)dB d:km
Noise power spectral density	−174.00 dBm/Hz
Number of slots	7
Data subcarrier of each resource block	12

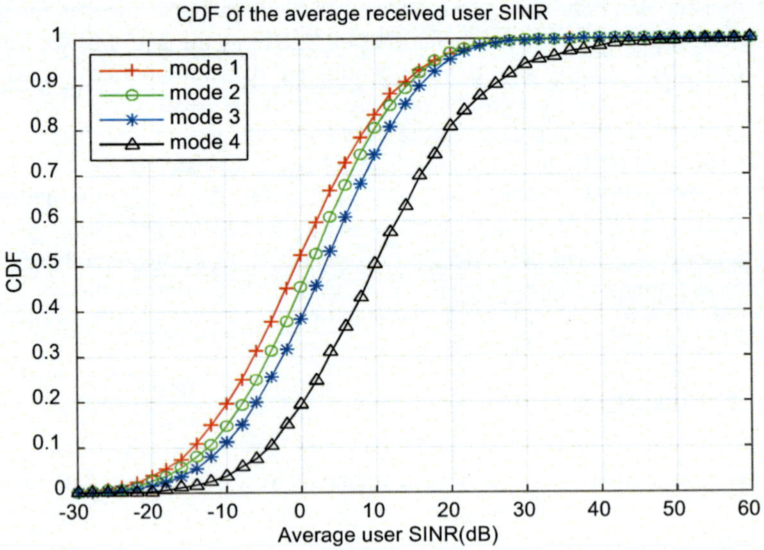

Fig. 4 SINR distribution curve of four modes

plays an important role in obtaining higher throughput and fairness. It determines the allocation method of wireless resources. Commonly used resource scheduling algorithms are maximum carrier-to-interference ratio algorithm (MAX C/I), polling algorithm (RR), and proportional fairness (PF) algorithm. Different resource allocation methods are validation in our simulations.

Firstly, four FR modes are simulated independently, and their SINR distribution curves are shown in Fig. 4. As can be seen from Fig. 4, mode (a) has the highest spectrum utilization, but the inter-cell interference is the most serious, and the coverage is the worst. If the number of edge users is large, mode (b) should be selected to achieve better performance. Mode (c) is appropriate if the user has a lot of traffic and wants to get a lot of throughput. If the inter-cell interference is minimum and the coverage area is large, mode (d) should be adopted.

The simulation results of FR based on RL are shown in Fig. 5. According to the change of cellular environment, one of the four FR modes will be used in the community. This proves that cellular can automatically select FR mode according to the change of environment with reinforcement learning-based decision-making. The effect of this method on network throughput is shown in Table 3. From Table 3, we can see that the network throughput of intelligent FR method is significantly higher than that of other FR methods, and it maintains a better coverage.

Finally, we simulate the signal interference characteristics based on the intelligent multimode FR method, and the results are shown in Figs. 6 and 7. The statistical level of each method is judged by different SINR. Cumulative distribution function (CDF) is not the same under the same SINR condition. From Figs. 6 and 7, we can

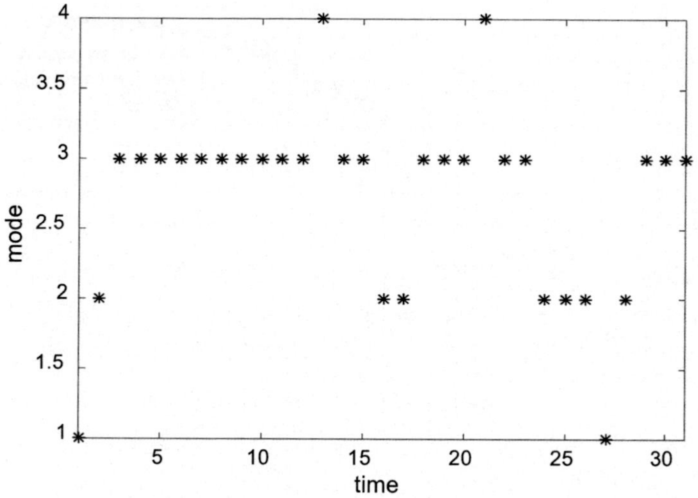

Fig. 5 The system switches between different modes

Table 3 Performance comparison of various FR methods

Mode name	Throughput (Mbps)	Coverage
Mode 1	12.8834	0.6751
Mode 2	14.3054	0.7456
Mode 3	16.9723	0.7986
Mode 4	9.8832	0.9562
Dynamic multimode frequency reuse	14.9742	0.8746

see that the SINR of our method for most edge users can be greater than 10 dB, while other methods are mostly less than 10 dB. Therefore, the SINR of all users and edge users has been significantly improved by using the FR method proposed in this chapter.

4 Conclusion

The existing frequency reuse algorithms have some shortcomings; that is, they cannot adjust the FR factor according to the change of the cell environment to achieve the goal of network performance optimization. In this chapter, an intelligent multimode FR method based on reinforcement learning is proposed to solve this problem, which can automatically select the most appropriate frequency reuse mode to improve the network throughput and coverage according to the real-time location and traffic situation of the users in the cell. The simulation results show that the FR method improves the throughput of the system and the coverage of the system.

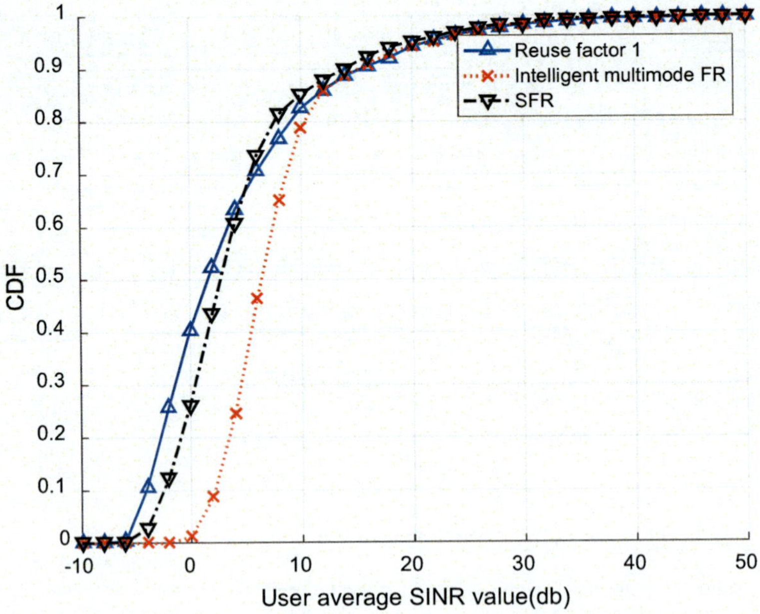

Fig. 6 Average SINR of all users

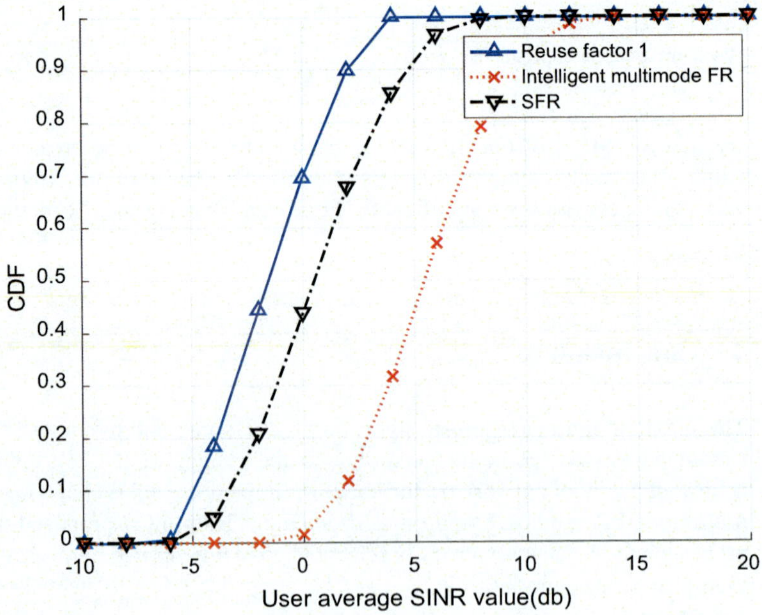

Fig. 7 Average SINR of edge users

References

1. L. Liu, T. Peng, P. Zhu, et al., Analytical Evaluation of Throughput and Coverage for FFR in OFDMA Cellular Network, in *IEEE Vehicular Technology Conference*. IEEE, (2016)
2. D. Lee, G.Y. Li, S. Tang, Intercell interference coordination for LTE systems. IEEE Trans. Veh. Technol. **62**(9), 4408–4420 (2013)
3. G. Boudreau, J. Panicker, N. Guo, et al., Interference coordination and cancellation for 4G networks. IEEE Commun. Mag. **47**(4), 74–81 (2009)
4. A. Simonsson, Frequency reuse and intercell interference coordination in E-UTRA, in *IEEE Vehicular Technology Conference, Dublin, Ireland*, (Sep. 2007), pp. 3091–3095
5. M. Rahman, H. Yanikomeroglu, W. Wong, Interference Avoidance with Dynamic Inter-Cell Coordination for Downlink LTE System, in *Wireless Communications & Networking Conference*, IEEE, (2009)
6. H. Mei, J. Bigham, P. Jiang, et al., Distributed dynamic frequency allocation in fractional frequency reused relay based cellular networks. IEEE Trans. Commun. **61**(4), 1327–1336 (2013)
7. A. Imran, M.A. Imran, R. Tafazolli, editors, A novel Self Organizing framework for adaptive Frequency Reuse and Development in future cellular networks. Personal Indoor and Mobile Radio Communications (PIMRC), in *2010 IEEE 21st International Symposium on*; (2010) 26–30 Sept. 2010
8. P. Zhu, T. Peng, Z. Liu, Energy Efficiency and Optimal Threshold for FFR chemes in OFDMA Cellular Networks, in *IEEE 10th International Conference on Signal Processing and Communication Systems*, (2016)
9. J.Y. Hwang, Y. Oh, Y. Han, Adaptive frequency reuse scheme for interference reduction in two-hop relay networks, in *Vehicular Technology Conference*. IEEE, (2010)
10. S. Alotaibi, R. Akl. Dynamic Fractional Frequency Reuse (FFR) Scheme for Two-Tier Network in LTE. in *IEEE 8th Annual Ubiquitous Computing, Electronics and Mobile Communication Conference*, (2017), pp. 408–413
11. X. Fangmin, Fractional Frequency Reuse (FFR) and FFR-Based Scheduling in OFDMA Systems, in *International Conference on Multimedia Technology*. IEEE, (2010)
12. G. Huang, J. Li, Interference mitigation for Femtocell networks via adaptive frequency reuse[J]. IEEE Trans. Veh. Technol. **65**(4), 2413–2423 (2016)
13. M. Qian, W. Hardjawana, Y. Li, Adaptive soft frequency reuse scheme for wireless cellular networks[J]. IEEE Trans. Veh. Technol. **64**(1), 118–131 (2015)
14. X. Yang, A multilevel soft frequency reuse technique for wireless communication systems. IEEE Commun. Lett. **18**(11), 1983–1986 (2014)
15. C. Yao, C. Yang, Z. Xiong, Energy-saving predictive resource planning and allocation. IEEE Trans. Commun. **64**(12), 5078–5095 (2016)
16. M. Belleschi, G. Fodor, D.D. Penda, Benchmarking practical RRM algorithms for D2D communications in LTE advanced. Wirel. Pers. Commun. **82**(2), 883–910 (2015)
17. M.N. Tehrani, M. Uysal, H. Yanikomeroglu, Device-to-device communication in 5G cellular networks: Challenges, solutions, and future directions. IEEE Commun. Mag. **52**(5), 86–92 (2014)
18. S. Deb, P. Monogioudis, J. Miernik, et al., Algorithms for enhanced inter-cell interference coordination (eICIC) in LTE HetNets. IEEE/ACM Trans. Networking **22**(1), 137–150 (2014)
19. L. Busoniu, R. Babuska, B.D. Schutter, D. Ernst, *Reinforcement Learning and Dynamic Programming Using Function Approximators* (CRC Press, Boca Raton, FL, 2010)

Image Registration Method for Temporal Subtraction Based on Salient Region Features

Suguru Sato, Huimin Lu, Hyoungseop Kim, Seiichi Murakami, Midori Ueno, Takashi Terasawa, and Takatoshi Aoki

1 Introduction

Cancer is the most common cause of death in Japan. It is reported that cancer causes 28.5% of death [1]. Bone is the most common site for metastasis in cancer and is of particular clinical importance in breast and prostate cancers because of the prevalence of these diseases. At postmortem examination, 70% of patients dying of these cancers have evidence of metastatic bone disease. Carcinomas of the thyroid, kidney, and bronchus also commonly give rise to bone metastases, with an incidence at postmortem examination of 30–40% [2]. Bone metastasis causes pain, pathologic fracture, hypercalcemia, myelosuppression, spinal cord compression, and nerve root lesions Spinal cord compression is urgent for diagnosis and treatment because it causes irreversible nerve palsy [3].

Computed tomography (CT) is frequently used as an image inspection method in the current medical field. It contributes to early detection of bone metastasis and evaluation of bone lesion size and bone stability. However, the number of images has increased due to the improvement of the device performance, which is a burden to the radiologist. In addition, inexperience radiologists have possibility of missing small and haze lesions. Computer-aided diagnosis (CAD) system is expected to contribute to solving these problems, and development has been advanced in recent years. The CAD system is a diagnostic system in which a doctor considers computer output such as detection and characterization of a lesion from a radiation image based on computer vision as a second opinion. It is only used as a tool for providing

S. Sato · H. Lu · H. Kim (✉)
School of Engineering, Kyushu Institute of Technology, Fukuoka, Japan
e-mail: kim@cntl.kyutech.ac.jp

S. Murakami · M. Ueno · T. Terasawa · T. Aoki
University of Occupational and Environmental Health, Fukuoka, Japan

Y. Li, H. Lu (eds.), *3rd EAI International Conference on Robotic Sensor Networks*,
EAI/Springer Innovations in Communication and Computing,
https://doi.org/10.1007/978-3-030-46032-7_2

additional information to doctors and does not have to be equal to or greater than the performance of a doctor. It needs to be complementary to the doctor's ability [4].

Temporal subtraction technique is one of the fundamental technologies of CAD system. It is a technique to generate images emphasizing temporal changes in lesion region by performing a difference calculation between current image and previous one of the same subject [5]. However, there is a problem that accurate temporal subtraction images cannot be created due to deviation of the photographing position by a difference in photographing time and body motion of the patient at the time of photographing. Image registration is necessary to solve this problem. There are many previous studies on image registration [6–8]. However, despite of the urgent need to diagnose bone metastasis, there are few studies on image registration focusing on vertebral region. Therefore, we propose an image registration system of the vertebral region in this chapter.

The proposed method consists of three steps. First, we perform segmentation of the region of interest (ROI) using position information of the spine based on biology. Next, we select pairs of previous image and current image in which same portion is depicted. Finally, we perform final image matching based on salient region features (SRF) [9]. We apply the proposed method to 65 cases. We confirmed the usefulness of temporal subtraction images for presenting the presence or absence of bone metastasis. Section 2 shows details of our proposed method. Section 3 shows the performance evaluation of the proposed method with synthetic data and an example of experimental results with real CT data. Finally, we conclude this chapter in Sect. 4.

2 The Proposed Method

The procedure of the proposed method is shown in Fig. 1. In our method, we performed noise removal by median filter and background removal using graph cut which is one of region segmentation method. Next, we select pairs of images by global image matching using center of gravity and normalized cross correlation (NCC) and perform final image matching based on SRF. For more details, please refer to the following.

2.1 Segmentation of Vertebral Region

We perform segmentation of the vertebral region as preprocessing for registration. From the anatomical point of view (i.e., vertebral region is located at center of the lower part on CT image), we set a rough region including the spinal region as a ROI. We apply median filter to ROI to eliminate noise.

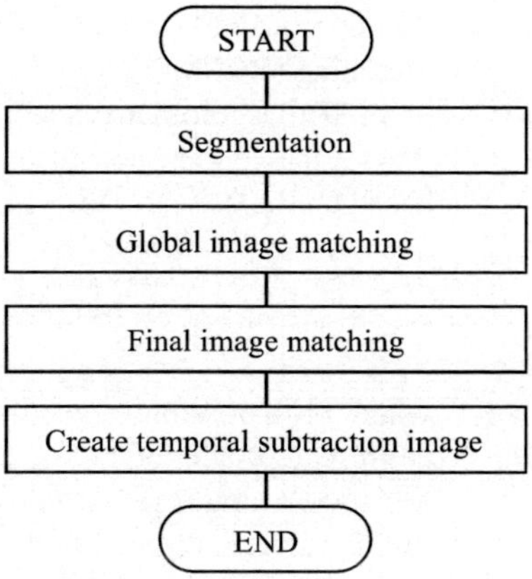

Fig. 1 The scheme of the proposed method

2.2 *Global Image Matching*

In order to detect the difference of the CT slice in the body axis direction, we select slices of previous image and current image of the same vertebral region is drown. We calculate the position of the center of gravity of previous and current first slice images. The center position on previous image $I_p(i,j)$ is translated in a rectangular area centered on the center of gravity position of current image $I_c(i,j)$, and calculate R_i that maximizes normalized cross correlation R_{NCC}.

$$R_{\mathrm{NCC}} = \frac{\sum\sum I_p(i,j)\, I_c(i,j)}{\sqrt{\sum\sum I_p(i,j)^2 \times \sum\sum I_p(i,j)^2}} \tag{1}$$

$$R_l = \max R_{\mathrm{NCC}} \tag{2}$$

We use the same operation for arbitrary number of previous images d while updating slices of current image one by one and calculate maximum normalized cross correlation $R_l(l = 1, 2, \cdots, d)$. The current image having the maximum value among calculated R_l is selected as first pair. A slice on current image selected as first pair is updated, and the maximum normalized cross correlation is calculated with several slices before and after first slice of previous image. After that, slice of current image is updated one by one, and all pairs are determined in the same procedure. In global image matching, we need to pay attention to the intersection of slice order when creating pairs.

2.3 *Final Image Matching*

We use conspicuity based on entropy and select SRF from vertebral region. Local image matching based on SRF consists of three steps: Salient Region Feature

Detection (SRFD), Region Component Matching (RCPM), and Region Configure Matching (RCFM).

In SRFD, automatic detection of SRF necessary for later processing is performed. We set the center coordinate X and the radius s. We calculate the entropy $H(s, X)$ based on the concentration probability function $p(s, X)$ in the set circular area.

$$H(s, X) = -\int_R p_i(s, X) \log_2 p_i(s, X)\, di \tag{3}$$

where i is a possible concentration value. After calculating the entropy at the center coordinate X and the radius s, the radius is updated at the same center coordinate X, and the entropy is recalculated. The radius is updated from 1 pixel to 10 pixels. The radius value with maximum entropy value is defined as optimum scale S_x. Feature value $A(S_x, X)$ is expressed by optimum scales.

$$A(S_x, X) = H \cdot S_x \cdot \int_R \left\| \frac{\partial}{\partial s} p_i | S_x \right\| di \tag{4}$$

The above process is the flow of SRFD at one center coordinate, and one feature value is calculated for one center coordinate. Then, next center coordinate on the image is set, and similar process is performed on the center coordinate. This is performed on the entire image. In this chapter, the updated interval of the center coordinates is set to 10 [pixel] in the X and Y directions. After calculating the feature value at each center coordinate, these feature values are compared, and top N number of circular regions is selected from the current image and the previous image, where $N = 20$.

Next in RCPM, after automatically determining the pair P of SRF between images based on similarity, we calculate the local likelihood $L_l(P)$ of the pair. We select one by one from the SRF of current image a and previous image b. We set these as pair P_{ab} and calculate the local likelihood $L_l(P_{ab})$ between selected SRF. The entropy correlation coefficient (ECC) which is normalized mutual information used for likelihood estimation is shown below:

$$L_l(P_{ab}) = \max_{\theta} \mathrm{ECC}(a, b_\theta) \tag{5}$$

$$\mathrm{ECC}(a, b_\theta) = 2 - \frac{2H(a, b_\theta)}{H(a) + H(b_\theta)} \tag{6}$$

where b_θ is a region b rotated at the rotation angle θ and H is binding entropy or peripheral entropy based on the density values of the two regions. When likelihood estimation is performed, region b is rotated every 3[deg] in the section of [−10, 10] to maximize ECC and select top M pairs, where $M = 30$.

In RCFM, combination of pairs is performed incrementally using the top M pairs selected by RCPM. We calculate global deformation amount T_g that transforms reference image for registration with target image based on local deformation amount T_l obtained from each pair. The movement amounts t_x and t_y are represented by displacement of the center coordinate of SRF. Rotation amount is shown below:

$$\hat{\theta} = \arg\max_{\theta'} \mathrm{ECC}\left(a, b_{\theta'}\right) \tag{7}$$

where $\hat{\theta}$ is the angle at which ECC obtains maximum value when SRF of the previous image is rotated by angle θ'. The likelihood L_g is used to evaluate the similarity of the entire image. This evaluates using ECC between target image I_F and image $I_{T'}$ obtained by transforming reference image I_T and measures likelihood of a pair combination composed by k pairs.

$$L_g\left(P_{a_1b_1} \cap \cdots \cap P_{a_kb_k}\right) = \mathrm{ECC}\left(I_F, I_{T'}\right) \tag{8}$$

When the pairs are incrementally combined, the pair showing maximum likelihood is selected as the pair to be selected first among M pairs. After selecting the first pair, pairs are incrementally combined so that the global likelihood is maximized. P'_{ab} is an additional candidate pair. $P_j = \left(P_{a_1b_1} \cap P_{a_2b_2} \cap \cdots \cap P_{a_kb_k}\right)$ is the latest pair combination. The global deformation amount T_g obtained from pair combination P_j is shown below:

$$T_g = \frac{1}{L_{\mathrm{sum}}} \sum_{i=1}^{k} L_{l_i} T_{l_i} \tag{9}$$

where L_{sum} is the sum of each pair constituting a pair combination, L_{l_i} is likelihood of the k-th pair of pair combination, and T_{l_i} is deformation amount of the k-th pair of pair combination. After the first pair is determined, we select a maximum likelihood pair from pairs not included in the latest pair combination as the additional candidate pair P'_{ab}. After a pair of additional candidate is selected, likelihoods are compared in the case of combining or not combining additional candidates. If the likelihood $L_g\left(P_j \cap P'_{ab}\right)$ obtained by combining additional candidates is larger than the likelihood $L_g(P_j)$ of current pair combination, additional candidate combinations are performed and pair combination is updated to $P_j = P_j \cap P'_{ab}$. If pairs of additional candidates that are not included in the latest pair combination exist after the pair combination has been updated, maximum likelihood pairs are selected from them, and same processing is performed. If likelihood $L_g\left(P_j \cap P'_{ab}\right)$ obtained by combining the additional candidates is lower than current likelihood $L_g(P_j)$ of pair combination, or there are no pairs of additional candidates which are not included in the latest pair combination, iterative process is terminated. The global deformation amount T_g is calculated using the latest pair combination, and reference image is deformed.

3 Experimental Results

In this chapter, we evaluated registration accuracy in vertebral region using synthesized data. An image of a certain vertebral region was defined as reference image, and an image transformed by applying processing to the reference image was defined as target image. The rate of overlapping region of the vertebral region of target and reference image was defined as true positive (TP), and the rate of not overlapping region (leakage region) was defined as false positive (FP). TP and FP are shown below:

$$\mathrm{TP} = \frac{n\,(A \cap B)}{n(A)} \times 100 \tag{10}$$

$$\mathrm{FP} = \frac{n(C)}{n(B)} \times 100 \tag{11}$$

where A is the spinal region of deformed target image, B is the spinal region of reference image, C is the leakage region, and n is the number of pixels.

We used four synthetic data as target images: (i) the image obtained by rotating reference image at rotation angle $\theta_a = -10$, (ii) the image obtained by applying the Gaussian filter to rotated image, (iii) the image obtained by adding artificial pseudo-lesion region to rotated image, and (iv) the image obtained by adding random noise of 5% to rotated image. Table 1 shows the performance evaluation, which is obtained TP and FP of each target image. Furthermore, we compare the result of matching by proposed method with the result of marking by radiologist.

Figure 2 shows three examples of matching result by proposed method and marking result by radiologist. In Fig. 2, (a), (d), and (g) show temporal subtraction images; (b), (e), and (h) show fusion images after matching; and (c), (f), and (i) show marking images of lesion by radiologist. In fusion image, reference image is shown in red, target image is shown in green, and overlapping region is shown in yellow color. From the experimental results, abnormal area is enhanced on the temporal subtraction images.

Table 1 Performance evaluation

	TP (%)	FP (%)
Rotation	100.0	12.16
Gaussian filter	70.40	0.00
Pseudo-lesion region	99.45	17.89
Random noise	83.05	16.95

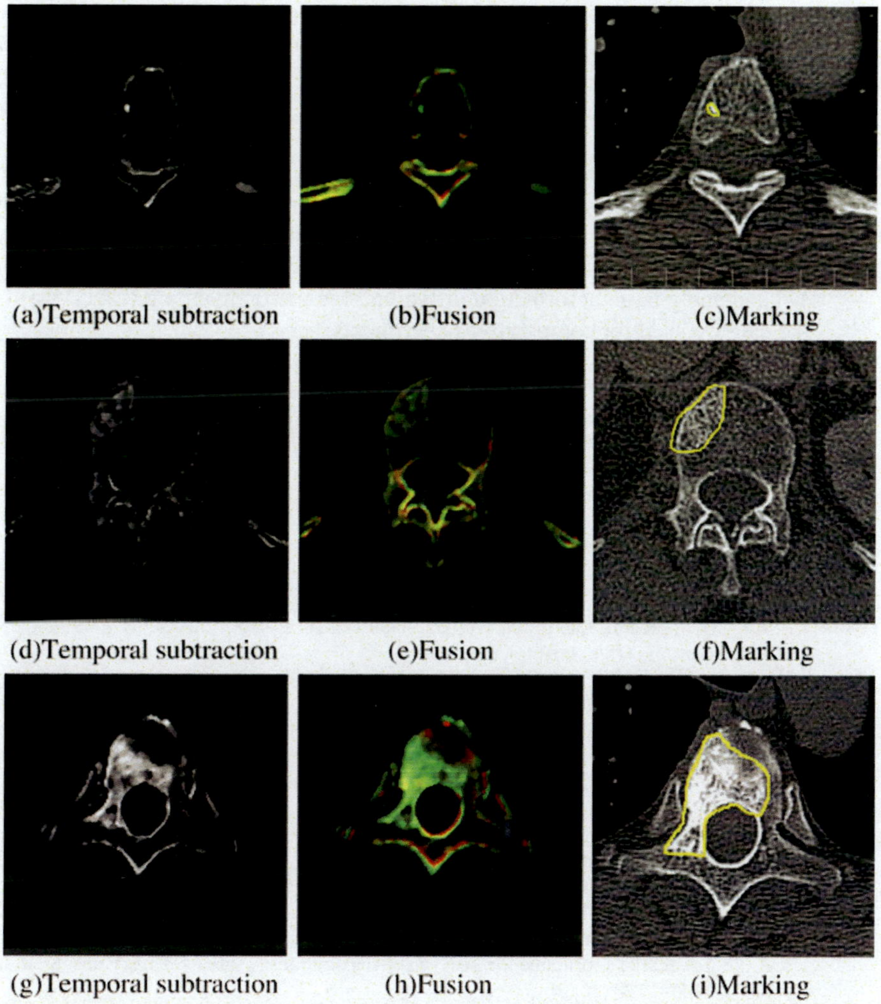

Fig. 2 Example of experimental result. (**a**) Temporal subtraction. (**b**) Fusion. (**c**) Marking. (**d**) Temporal subtraction. (**e**) Fusion. (**f**) Marking. (**g**) Temporal subtraction. (**h**) Fusion. (**i**) Marking

4 Conclusion

In this chapter, we proposed a method of image registration between current and previous images in the vertebral region as a preliminary study to generate temporal subtraction images from CT images and detected bone metastases on the vertebral region. We evaluated performance using synthetic data. The image obtained by rotating reference image at rotation angle $\theta\_a = -10°$ gave TP 100.0% and FP 12.16%. The image obtained by applying the Gaussian filter to rotate image gave

TP 70.40% and FP 0.00%. The image obtained by adding artificial pseudo-lesion region to rotated image gave TP 99.45% and FP 17.89%. The image obtained by adding random noise of 5% to rotated image gave TP 83.05% and FP 16.95%.

The proposed method succeeded in emphasizing the temporal changes and contributed to improving the detection accuracy of the lesion part of doctor. This deviation is due to the slice of same place that is not selected by global matching, and only local deformation amount of the first pair is used as the global deformation amount in RCFM step. In the future work, in order to improve the effectiveness of the proposed method, we need to apply optimization using swarm intelligence [10, 11] and apply large deformation diffeomorphic metric mapping (LDDMM) [12] to the data we used for comparative experiments.

Acknowledgement This work was supported by Leading Initiative for Excellent Young Researchers of the Ministry of Education, Culture, Sports, Science and Technology-Japan (16809746) and JSPS KAKENHI Grant Number 16K14279, 17K10420.

References

1. Ministry of Health, Labor and Welfare, Vital statistics, http://www.mhlw.go.jp/toukei/saikin/hw/jinkou/suikei16/index.html
2. R.E. Coleman, Clinical features of metastatic bone disease and risk of skeletal morbidity. Clin. Cancer Res. **12**(20), 6243–6249 (2006)
3. G.J. Cook et al., Detection of bone metastases in breast cancer by FDG PET: Differing metabolic activity in osteoblastic and osteolytic lesions. J. Clin. Oncol. **16**(10), 3375–3379 (1998)
4. K. Doi, Computer-aided diagnosis in medical imaging: Historical review, current status and future potential. Comput. Med. Imaging Graph. **31**(4–5), 198–211 (2007)
5. K. Doi et al., Computer-aided diagnosis in radiology: potential and pitfalls. Eur. J. Radiol. **31**(2), 97–109 (1999)
6. Z. Zhong et al., 3D-2D deformable image registration using feature-based nonuniform meshes. BioMed Res. Int. **2016** (2016)
7. A. Sotiras et al., Deformable medical image registration: A survey. IEEE Trans. Med. Imaging **32**(7), 1153–1190 (2013)
8. B. Rodriguez-Vila et al., Methodology for registration of distended rectums in pelvic CT studies. Med. Phys. **39**(10), 6351–6359 (2012)
9. X. Huang et al., Hybrid image registration based on configural matching of scale invariant salient region features. CCVPRW **11**, 167–179 (2004)
10. D. Karaboga, An Idea Based on Honey Bee Swarm for Numerical Optimization, Technical Report-TR06, Erciyes University, Engineering Faculty, Computer Engineering Department, (2005)
11. X.S. Yang, Cuckoo Search via Levy Flights, in *World Congress on Nature & Biologically Inspired Computing*, IEEE Publications, (2009), pp. 210–214, arXiv:1003.1594v1
12. R. Sakamoto, Temporal subtraction of serial CT images with large deformation diffeomorphic metric mapping in the identification of bone metastases. Radiology **285**(2), –629, 639 (2017)

Extreme ROS Reality: A Representation Framework for Robots Using Image Dehazing and VR

Akira Ueda, Huimin Lu, and Hyoungseop Kim

1 Introduction

Japan is known as a country that is likely to suffer natural disasters such as typhoons, heavy rains, heavy snow, floods, landslides, earthquakes, tsunamis, and volcanic eruptions worldwide. Although the area of Japan's land is only 0.28%, 20.5% of earthquakes of magnitude 6 or more that happened in the whole world occurred in Japan, and 7.0% of active volcanoes in the whole world exist in Japan. In addition, 0.3% of disaster deaths worldwide is recorded in Japan, and 11.9% of the damage amount received in the disasters of the whole world is the damage amount in Japan [1]. Japan is more likely to suffer from natural disasters than other countries. In addition, artificial disasters such as accidents at construction sites (e.g., gas explosion accidents in tunnels) are also increasing. The deterioration of infrastructure such as tunnels, expressways, and dams is a problem, and complex accidents caused by aging infrastructure such as corrosion are on the rise. Responding to such equipment accidents is also a subject [2].

As the threat of natural disasters and artificial disasters increases, the importance of preparing for them increases. In terms of disaster prevention, it is important not only to take measures after the disaster such as emergency response, recovery, reconstruction, measures to prevent disasters, and inspection and maintenance of infrastructure and equipment [2].

It is always dangerous for a person to work in disaster response and recovery where it is difficult to do work. Disaster prevention measures and inspection and maintenance of social infrastructure and equipment are similar. In order to solve these problems, introduction of robots under dangerous work environments is

A. Ueda · H. Lu (✉) · H. Kim
School of Engineering, Kyushu Institute of Technology, Fukuoka, Japan
e-mail: dr.huimin.lu@ieee.org

Y. Li, H. Lu (eds.), *3rd EAI International Conference on Robotic Sensor Networks*, EAI/Springer Innovations in Communication and Computing,
https://doi.org/10.1007/978-3-030-46032-7_3

progressing. By employing robots on behalf of people, it is possible to drastically reduce the risk of injuries and deaths due to misoperation accidents.

Regarding robots at the disaster site, it is necessary to deal with various situations. Complete autonomy of the robot is required, and remote control technology is also essential. The mainstream of current remote control technology of robot is acquiring information on the working environment of the robot with a single sensor or a plurality of sensors and transmitting the obtained information to a monitor in a remote place where the operator is two-dimensional or three-dimensional data. The operator performs operations while recognizing the work environment of the robot from the monitor. In this method, since it is necessary to grasp the work environment of the robot from the monitor, the degree of difficulty is higher than the actual operation at the work site, and there is concern that the work accuracy and the efficiency decrease. For this reason, sufficient training is required to perform remote control of the robot, and only the pilot who has proficient skills is allowed to perform the operation.

Also, depending on the disaster site, there are many works under low light source without external illumination such as indoors. In order to acquire information on the surrounding environment using the camera, it is necessary to use light attached to the robot. In this case, unlike ceiling lighting which uniformly illuminates the surrounding environment, it is expected that clear images cannot be obtained because illumination attached to the robot illuminates the periphery inhomogeneously. Generally, when capturing an image in a dark environment such as nighttime or indoors, image quality deteriorates due to a decrease in noise and contrast. It will affect the image precessing such as object detection under such circumstances [3, 4].

In addition to the above, the current robot remote control technology is considered to have room for improvement due to various factors, and development of a new robot remote control system is required. In this chapter, for the purpose of supporting the development of a remote control system of a new robot, we will develop a method to reproduce working environment of robot using virtual reality technology and vision improvement technology of low light source image.

2 Image Dehazing

2.1 Principle

The method using the haze removal model is to adapt the haze removal method based on the fact that the inversion of the RGB components of the dark image is similar to haze such as fog.

As described in [6], in the images with haze such as fog or cloudiness, the background pixel value is always high for all channels. However, due to the influence of color vividness or shadow, in an area including main objects such as

Fig. 1 Example of low illumination image and its reversed image and haze image ([8, 9] data set)

houses, cars, and people, it takes a low value in at least one channel among the RGB channels. On the other hand, the minimum value of the RGB component is taken for the video taken under the low light source, and the similarity with the haze image is investigated. As a result, through 30 randomly chosen haze videos and RGB inverted low illumination images, research results have been obtained that the histograms of the minimum values of the RGB components of all pixels are very similar [7]. Figure 1 shows an image taken at night, an image after RGB component inversion operation, and a haze image including fog.

Based on the above observation, we applied a reversing operation of RGB components to low illumination image, applied haze elimination algorithm, then reversed the RGB components again, and proposed a low illumination image emphasis method as the final output was done. In this chapter, we use the method of document [5] using the above algorithm.

2.2 *Dehazing Model*

First, reverse processing is performed on the low illumination input image as follows.

$$I_{\mathrm{inv}}^{C}(x) = 255 - I^{C}(x) \tag{1}$$

where C is the RGB color channel. $I^C(x)$ is the input image captured under a low light source, and $I^C_{\text{inv}}(x)$ is the inverted image of the input image.

The image degradation model proposed in [15] is expressed as follows:

$$I_{\text{inv}} = J_{\text{inv}}(x)t(x) + A\,(1 - t(x)) \tag{2}$$

where $J_{\text{inv}}(x)$ is the luminance of the scene, A is the ambient light, and its ratio is determined by the transmittance $t(x)$. By finding A and $t(x)$ from the relationship expressed in Eq. (2), $J_{\text{inv}}(x)$ can be obtained from I_{inv} as follows:

$$J_{\text{inv}}(x) = \frac{I_{\text{inv}} - A}{t} + A \tag{3}$$

Finally, $J_{\text{inv}}(x)$ is converted and an image is output. In this process, it is necessary to estimate ambient light A and transmittance $t(x)$. The method of estimating these parameters and the method of obtaining the final improved image are described below.

2.3 Ambient Light A Estimation

First, as shown in the following equation, the minimum value of the RGB components is taken for the input image after the inversion processing.

$$I_{\min}(x) = \min_{C\in\{r,g,b\}}\left(I^C_{\text{inv}}(x)\right) \tag{4}$$

Since $I_{\min}(x)$ always includes a lot of noise, smoothing processing using a median filter is performed here. Next, the pixel with the highest brightness among the smoothed $I_{\min}(x)$ is selected, and the pixel value at that position is set as ambient light A.

2.4 Transmittance t Estimation

After estimating the ambient light, estimating the transmittance is needed. Estimate to maintain the transmittance $t(x)$ locally as smoothly as possible. First, the RGB minimum value is taken for the input image $I^C_{\text{inv}}(x)$ after inversion processing. This grayscale image is defined as a dark channel $\tilde{I}_{\text{dark}}(x)$.

$$\tilde{I}_{\text{dark}}(x) = \min_{C\in\{r,g,b\}}\left(\frac{I^C_{\text{inv}}(x)}{A^C}\right) \tag{5}$$

where Gaussian pyramid processing is applied for smoothing $\widetilde{I}_{\mathrm{dark}}(x)$. In this chapter, downsampling and upsampling are performed once. Then, a median filter is applied.

$$I_{\mathrm{med}}(x) = \mathrm{Med}_S\left(\widetilde{I}_{\mathrm{dark}}(x)\right) \tag{6}$$

S is the scale of the median filter. In order to smooth the transmittance while holding the edge, the following processing is performed.

$$I_{\mathrm{datail}}(x) = \mathrm{Med}_S\left(\left|I_{\mathrm{med}}(x) - \widetilde{I}_{\mathrm{dark}}(x)\right|\right) \tag{7}$$

$$I_{\mathrm{smooth}}(x) = I_{\mathrm{med}}(x) - I_{\mathrm{datail}}(x) \tag{8}$$

Furthermore, the dark channel can be optimized as follows:

$$I_{\mathrm{dark}}(x) = \begin{cases} \mu\widetilde{I}_{\mathrm{dark}}(x) & \text{if } \mu\widetilde{I}_{\mathrm{dark}}(x) < I_{\mathrm{smooth}}(x) \\ I_{\mathrm{smooth}}(x) & \text{others} \end{cases} \tag{9}$$

where $\mu\widetilde{I}_{\mathrm{dark}}(x)$ is set when $\mu\widetilde{I}_{\mathrm{dark}}(x)$ takes a value smaller than $I_{\mathrm{smooth}}(x)$, and $I_{\mathrm{smooth}}(x)$ is set as the value of dark channel in other cases. In this chapter, the value of μ is set to 0.95.

Then, the transmittance is obtained as follows:

$$t(x) = 1.0 - \omega I_{\mathrm{dark}}(x) \tag{10}$$

where ω is a parameter for adjusting the value of $t(x)$, which is set to 0.98 in this chapter.

2.5 *Dehazed Image*

$J_{\mathrm{inv}}(x)$ can be obtained by adapting the estimated transmittance $t(x)$ and ambient light A to Eq. (3). The RGB component is inverted by the following equation and output as the final improved image.

$$J^C(x) = 255 - J^C_{\mathrm{inv}}(x) \tag{11}$$

Figure 2 shows the results of the above processing for images taken under a low light source.

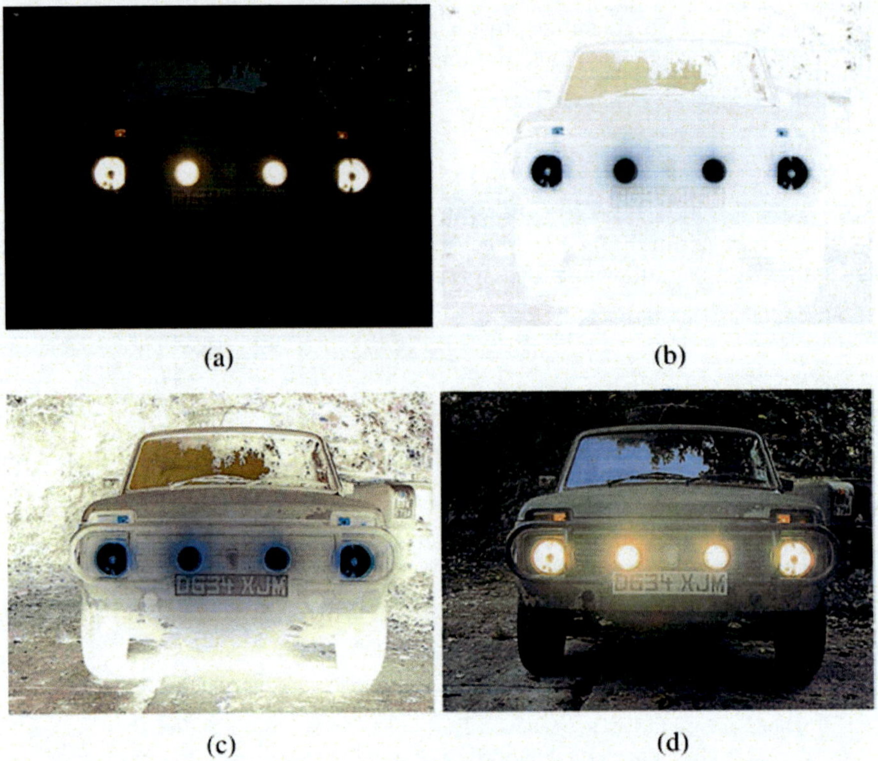

Fig. 2 Result of low light source image improvement process (data set of [8]). (**a**) Input image; (**b**) inversed image; (**c**) dehazed image; (**d**) output result

3 Virtual Reality-Based Robot Operation

3.1 Virtual Reality

In recent years, remote control technologies that operate the robot via a two-dimensional interface such as a monitor are spreading [10]. In these remote control technologies, 2D monitors and keyboards have been used to control the robot [11]. However, it is not easy to observe two-dimensional or three-dimensional information from a 2D monitor and to steer the robot using input devices such as a keyboard, and related knowledge and high control skills are required for remote operation. Therefore, in order to perform the remote control of the robot, it is necessary to receive sufficient training, and only the limited technician can perform the operation.

A new technique for simplifying the remote operation of the robot is indispensable. Therefore, in this chapter, instead of using the 2D monitor interface, we develop a method that uses virtual reality (VR) technology which has been showing remarkable success in various fields in recent years. By using the VR device for reproducing the work environment of the robot, it is possible to easily recognize the working environment. It is also thought that even a person who does not have the maneuvering technique will be able to remotely operate the robot only with simple training.

However, the current virtual reality system does not support frameworks that make up a robot system such as ROS (Robot Operating System) [12], and there is no standard interface for combining robot and VR system. Therefore, in this chapter, we refer to the ROS Reality [13] system developed by David Whitney and others and combine the robot system and the virtual reality system to reproduce the working environment of the remote place.

3.2 System Design

In order to construct a robot remote control system, it is necessary to first measure the robot work environment. In this chapter, we measure the working environment by acquiring color images and depth images using RGBD sensors. In the dark environment, image quality is improved for color images by the image analysis method described in Sect. 2. The obtained image data is transmitted to the computer executing the Unity game engine connected with the VR device via Rosbridge WebSocket [14]. Create point cloud data based on color image and depth image received on Unity computer and reproduce working environment of remote area. Figure 3 shows a schematic diagram of the remote work environment reproduction system described above. In addition, the details of the system will be described below.

3.3 Environment Restriction Result

Figure 4 shows an example of RGB image and depth image acquired. The result of reproducing on the Unity scene based on the obtained image information is shown in Fig. 5.

4 Experimental Results and Discussions

In this section, we describe the experimental environment and experimental results used in this chapter.

4.1 Devices

In this thesis, Kinect for Windows v2 (Kinect v2) was used to measure the work environment of the robot. Kinect v2 is a generic name for Kinect sensor and SDK

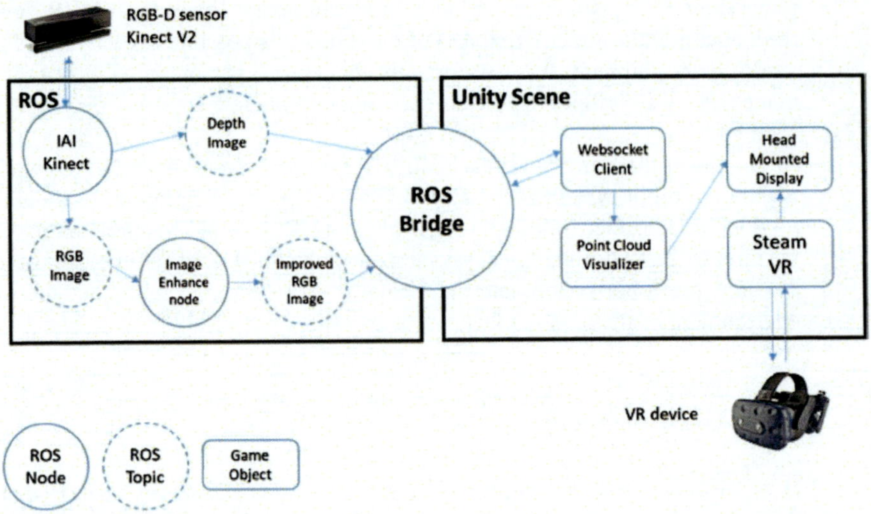

Fig. 3 The proposed system

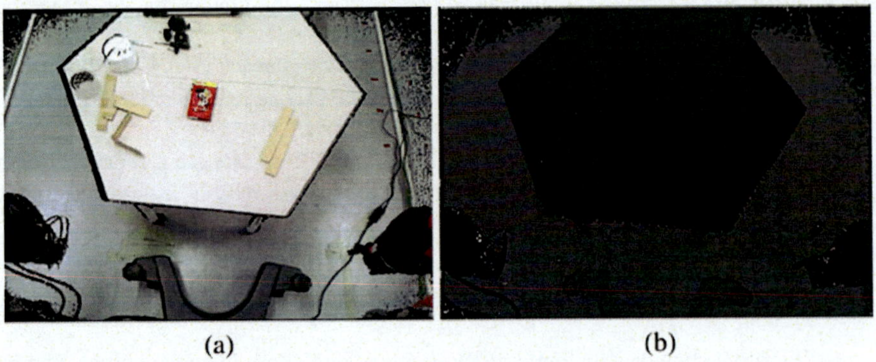

Fig. 4 Image information from Kinect. (**a**) RGB image; (**b**) depth image

released by Microsoft Corporation in 2014 for measuring color images and depth images. For the depth information, use the time-of-flight measurement method. The appearance of Kinect v2 is shown in Fig. 5, and the operation specification is shown in Table 1.

In this chapter, VIVE Pro Head Mounted Display (HMD) was used as an interface for working environment recognition. VIVE Pro HMD is one of the Virtual Reality Head Mounted Display products released by HTC Corporation.

In this chapter, we used two PCs to measure the work environment of the robot and to reproduce the work environment in the remote place. The specifications of each PC are shown in Fig. 6.

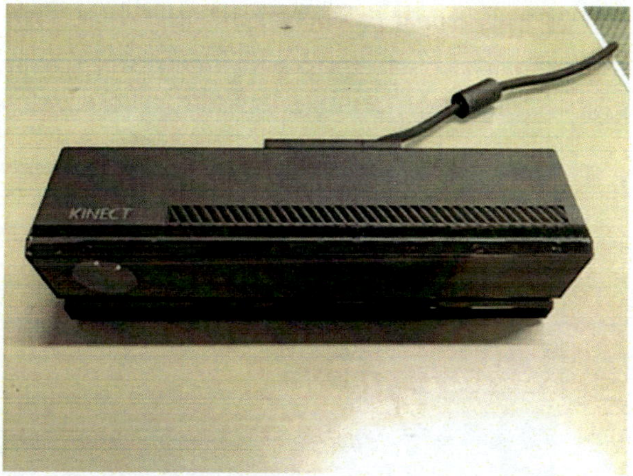

Fig. 5 KINECT imaging sensor

Table 1 Kinect v2

Color image	1920 × 1080 [pixel]
Frame rate of CI	30 [fps]
Depth image	512 × 424 [pixel]
Frame rate of DI	30 [fps]
Imaging type	ToF (time of flight)
Recognition range	500–8000 [mm]
Horizontal angle	70°
Vertical angle	60°

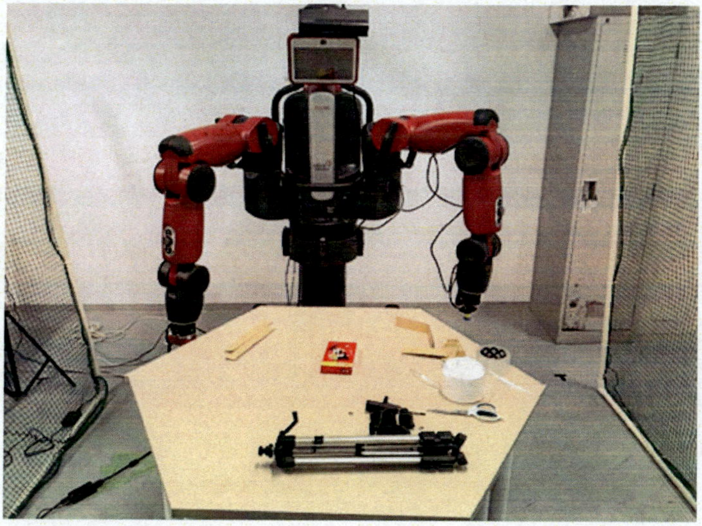

Fig. 6 The experimental environments

4.2 Experimental Results

We confirmed the utility of remote reality system using the virtual reality technology proposed in this chapter and image improvement processing under low light source. In this chapter, we measured work environments under low light sources and reproduced them in virtual reality space. At that time, a visual comparison was made between the case where the image was reproduced as it was based on the obtained image data and the case where the image improvement processing was performed and reproduced (Figs. 7 and 8).

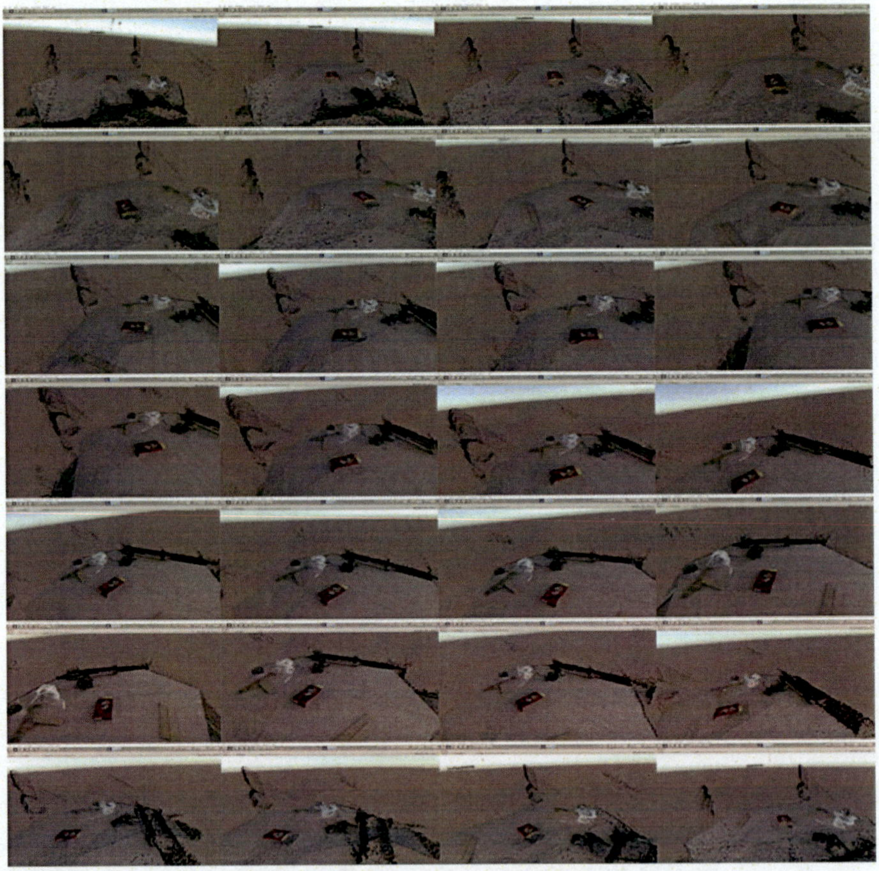

Fig. 7 Representation result without image enhancement

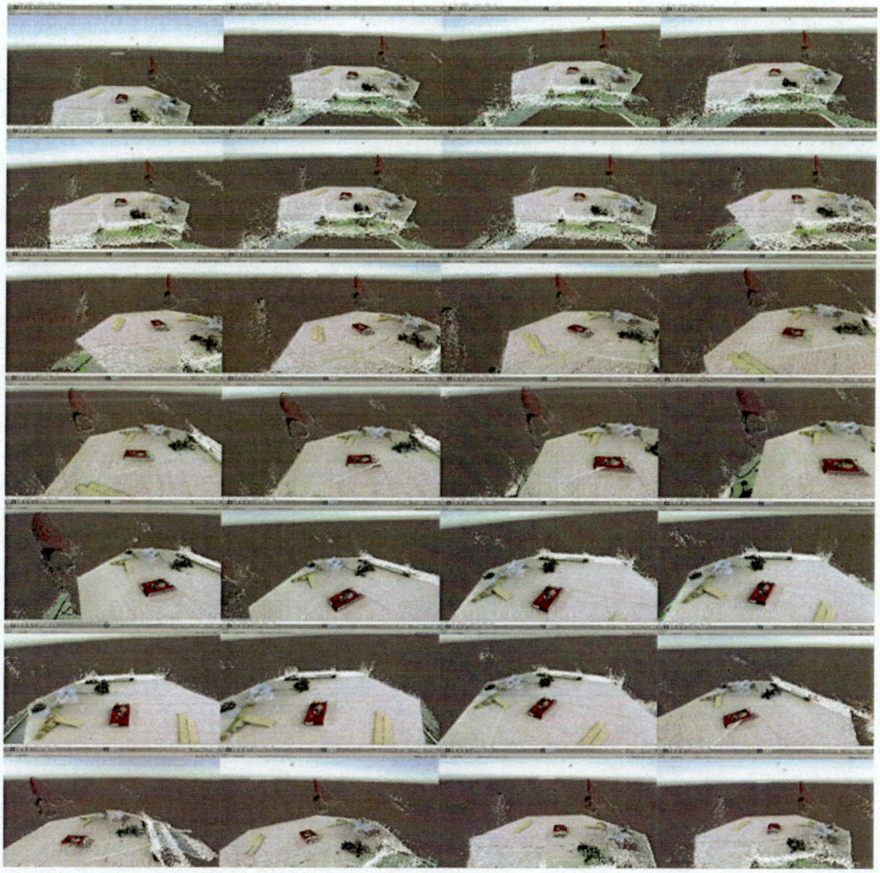

Fig. 8 Representation result with image enhancement

5 Conclusion

In this chapter, we proposed a method to reproduce working environment of robot for supporting development of remote operation system of new robot. In order to improve the visibility under the dark environment, the image improvement processing using the haze removal model was performed, and the working environment of the remote place was reproduced using the virtual reality technology. Image quality improvement processing is performed in a plurality of different dark environments, and comparison was made with the reproduction results in the case where the reproduction to the virtual reality space is performed and the case where the image improvement processing is not performed to improve the visibility by the improvement processing.

References

1. http://www.jice.or.jp/knowledge/japan/commentary09#jump_02. Accessed 30 Dec 2018
2. H. Lu, Y. Li, M. Chen, H. Kim, S. Serikawa, Brain intelligence: Go beyond artificial intelligence. Mobile Networks and Applications **23**, 368–375 (2018)
3. H. Lu, D. Wang, Y. Li, J. Li, X. Li, H. Kim, S. Serikawa, I. Humar, CONet: A cognitive ocean network. IEEE Wirel. Commun. **26**(3), 90–96 (2019)
4. K. Huang et al., A real-time object detecting and tracking system for outdoor night surveillance. Pattern Recogn. **41**, 432–444 (2018)
5. X. Jiang et al., Night Video Enhancement Using Improved Dark Channel Prior, in *IEEE International Conference on Image Processing*, (2013), pp. 553–557
6. K. He et al., Single image haze removal using Dark Channel prior. IEEE Trans. Pattern Anal. Mach. Intell. **33**, 2341–2353 (2011)
7. X. Dong et al., Fast efficient algorithm for enhancement of low lighting video. IEEE Int. Conf. Multimed. Expo, 1–6 (2011)
8. Y.P. Loh et al., Getting to know low-light with the exclusively dark dataset. Comput. Vis. Image Underst. **178**, 30–42 (2019)
9. K. Ma et al., Perceptual Evaluation of Single Image Dehazing Algorithms, in *IEEE International Conference on Image Processing*, (2015), pp. 3600–3604
10. K. Goldberg et al., Desktop teleoperation via the world wide web. IEEE Int. Conf. Robot. Autom. **1**, 654–659 (1995)
11. J. Vertut, *Teleoperation and Robotics: Applications and Technology* (Springer Science & Business Media, London, 2013)
12. Robot Operating System, http://wiki.ros.org/ja
13. D. Whitney et al., ROS Reality: A Virtual Reality Framework Using Consumer-Grade Hardware for ROS-Enabled Robots, in *IEEE/RSJ International Conference on Intelligent Robots and Systems (IROS)*, (2018), pp. 1–9
14. C. Crick et al., Rosbridge: ROS for Non-ROS users. Robot. Res., 493–504 (2017)
15. H. Lu, Y. Li, L. Zhang, S. Serikawa, Contrast enhancement for images in turbid water. J. Opt. Soc. Am. A Opt. Image Sci. Vis. **32**(5), 886–893 (2015)

Double-Blinded Finder: A Two-Side Privacy-Preserving Approach for Finding Missing Children

Xin Jin, Shiming Ge, Chenggen Song, Xiaodong Li, Jicheng Lei, Chuanqiang Wu, and Haoyang Yu

1 Introduction

The impact of missing children on the vast majority of families is immeasurable. According to reports, there were 460,000 people in the USA, the United Kingdom, and Germany, and 112,853 and 100,000 children were missing. An effective way to get back the missing child is to use a social network that involves both parties. The Social Network Helpers (SNH) take photos of the suspicious missing photos in public places and upload them to a public cloud server. Then the Parent(s) of Missing Children (PMC) can upload the photo of his/her missing children and then automatically search via face matching. The results are reported to the Center

X. Jin
Beijing Electronic Science and Technology Institute, Beijing, China

State Key Laboratory of Virtual Reality Technology and Systems, Beihang University, Beijing, China

S. Ge (✉)
Institute of Information Engineering, Chinese Academy of Sciences, Beijing, China
e-mail: geshiming@iie.ac.cn

C. Song
OracleChain Technology, Beijing, China

X. Li
Beijing Electronic Science and Technology Institute, Beijing, China

CETC Big Data Research Institute Co., Ltd., Guizhou, China

J. Lei · C. Wu
CETC Big Data Research Institute Co., Ltd., Guizhou, China

H. Yu
Beijing Electronic Science and Technology Institute, Beijing, China

Y. Li, H. Lu (eds.), *3rd EAI International Conference on Robotic Sensor Networks*, EAI/Springer Innovations in Communication and Computing,
https://doi.org/10.1007/978-3-030-46032-7_4

of Missing & Exploited Children (CME) if the true missing children are found. Due to the increasing number of children's photo exposures, this can lead to many dangerous problems and security risks **privacy**.

The **effectiveness** of detection and identification is also a very challenging part of the field of blind computing. There are three main reasons for this: (1) Due to factors such as the direction of the face in the street, illumination, and occlusion, it is very difficult to perform accurate detection in the case of blind calculation. (2) Children's *robust representation* is also an important reason for the accuracy of recognition. (3) *Effective matching* of facial features should also meet the requirements of blind computing. To address privacy and effectiveness issues, some works have been proposed in blind computation and face recognition [4–7, 12].

Inspired by that, this work proposes Double-Blinded Finder, a social network system for finding missing children in a double-blinded way. Unlike the existing social networks which usually contain two parties, Double-Blinded Finder involves three parties (SNH, PMC, and CME) and mainly consists of face representation module in mobile clients and face matching module in the cloud, as shown in Fig. 1. In the mobile client, face representation module applies an easy-to-use user interface to facilitate child face detection, representation, and confirmation. Each detected child face is represented as a 128d feature vector and two auxiliary attributes (gender and age) with a deep face recognizer, MT-FaceNet, which is trained on our built Labeled Child Face in the Wild (LCFW) dataset in a multi-task manner. Then, SNH upload encrypted photos of suspicious missing children to the public cloud using the public keys generated by the face feature vector and the public parameters from CME. PMC first send a feature vector extracted from the photo of his/her missing child to CME. Then a pair of private keys is sent back to PMC. Finally, PMC send one of the private key to the public cloud for matching faces in the cloud using our proposed blind face matching algorithm. The matched encrypted face is sent to PMC and decrypted using the other private key, and then confirmed by PMC.

In the cloud server, an effective face blind identification matching method is proposed. The face photo from the SNH and the face representation from the PMC are encrypted using the CME generation key. Then, blind face matching runs in a public cloud to avoid the leakage of the privacy of both sides by using inner-product encryption (IPE). The main contributions of this work are three folds:

- For the first time, a lost child discovery system that combines face matching accuracy and privacy protection is proposed, which can effectively protect children who are suspicious or truly lost.
- A Labeled Child Face in the Wild (LCFW) dataset is constructed, including 60K facial images with 6K unique identities, and annotated with different types of attributes for each child face. Based on this dataset, we train a multi-task FaceNet (MT-FaceNet) model to describe a child face as a 128d fixed-point feature vector as well as auxiliary attributes.
- Face-to-face matching can be safely run on the public cloud due to the face matching method based on IPE-based blind computing to restrict access to photos.

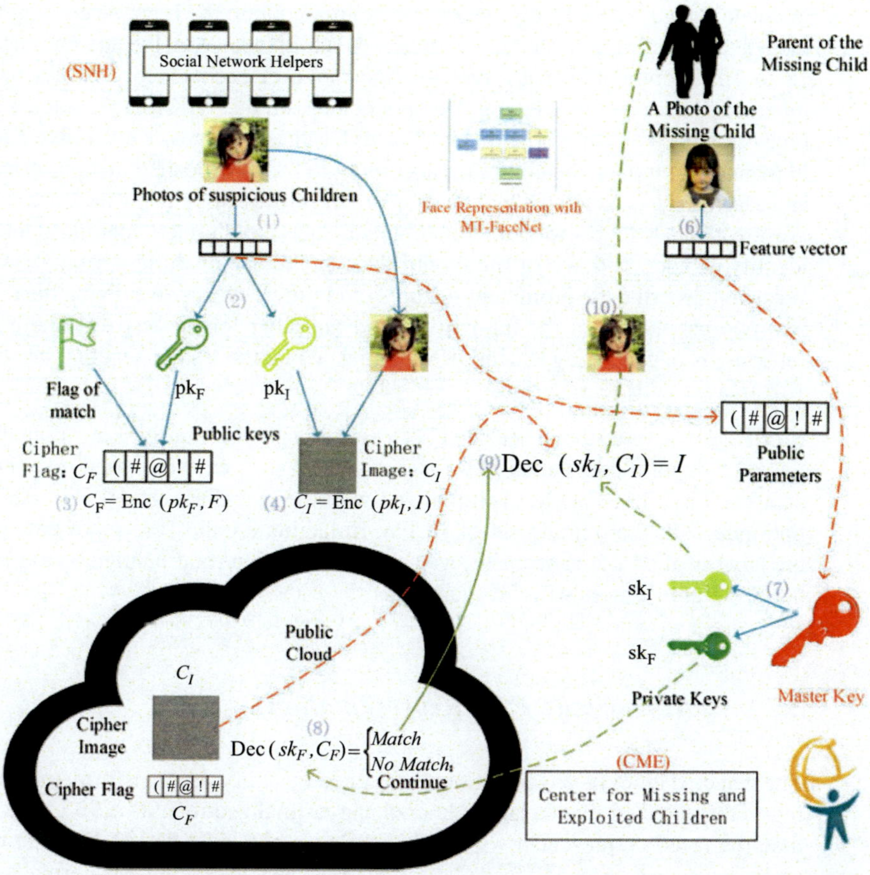

Fig. 1 Overview of Double-Blinded Finder. (1) Detect child faces from the photos of suspicious missing children taken by SNH and represent each child face as a feature vector. (2) Generate a public keys pair (pk_F and pk_I) from the feature vector and the public parameters of CME. (3) Encrypt a match flag F to cipher flag C_F with pk_F. (4) Encrypt the image taken by SNH to cipher image C_I with pk_I. (5) Upload C_F and C_I to the public cloud. (6) Detect and represent the child face from a photo of the true missing child. (7) Generate a private key pair (sk_F and sk_I) from the feature vector and the master key. (8) Compare sk_F with each cipher flag in the cloud by blind face matching. (9) Send C_I to PMC and decrypt it with sk_I if the query result matches. (10) Expose the decrypted image I to PMC for checking

2 Face Description via Deep Learning

2.1 Dataset, Model, and User Interface

Dataset We downloaded more than 300K images of more than 6K from a professional photo website to build our LCFW dataset, and used manual culling to delete images of children's faces that did not contain the logo. According to the metadata of the images, we only maintain the child faces with the age ranging from 18 to 108 months and make sure the age range of each identity is less than 60 months. Under careful selection, we obtained 6K unique 60K images, each with 10 images, each with a designated child face. On these images, we ask five subjects to manually annotate the child face according to the downloaded meta information, and each image is annotated by two subjects and cross-validated by the third subject. We use four attributes to manually annotate each child's face, including location, identity, gender, and age of the month. Due to the use of these attributes, we can very well describe the prominent features of children's faces. We divide the data set into two parts, one for the training set and the other for the test set. The training set consists of 50K images with 5K identities, while the testing set contains the rest 10K images with 1K identities.

Model We applied the FaceNet [11] extension to the multitasking domain and trained a depth model called MT-FaceNet, which uses datasets based on annotated datasets. First, MT-FaceNet is initialized with the OpenFace model [1].[1] Then, we fine-tune with the training set of LCFW to facilitate child face representation. In the process of data enhancement, we use the translation/rotation method to generate 500K children's faces with 5K identities.

2.2 *Multi-Attribute Face Representation*

The detected child face is first globally aligned by similarity transformation according to the detected five facial landmarks and normalized to a 96×96 image. After that, the face is represented with multi-attributes by using the MT-FaceNet model, including gender, age, and face feature. Finally, each face feature is normalized and quantized into 5-bits fixed point to meet the blind matching computation. Thus, a child face can be described as a 131d vector consisting of a 2d gender vector $\boldsymbol{h}$, a 1d age scalar s, and a 128d fixed-point identity vector $\boldsymbol{b}$.

$$\mathbf{f} = (\boldsymbol{h},\ s,\ \boldsymbol{b}) \tag{1}$$

In open mode, the pattern for face matching is to compare two identity vectors with age and gender constraints, representing a pair of face descriptors as matches according to the following:

$$\mathcal{M}(\mathbf{f}_1, \mathbf{f}_2) = \begin{cases} 1 & \boldsymbol{h}_1 \cdot \boldsymbol{h}_2 > t_g,\ |s_1 - s_2| < y_a,\ \boldsymbol{b}_1 \cdot \boldsymbol{b}_2 > y_s \\ 0 & \text{otherwise} \end{cases} \tag{2}$$

[1] http://cmusatyalab.github.io/openface/.

here, t_g, t_a, and t_s are three thresholds for restricting gender similarity, age difference, and identity similarity score, respectively.

2.3 Face Confirmation via Feedback

The facial image of the child from the PMC is encrypted and uploaded to form a query, while the facial image of the child from the SNH is encrypted and uploaded to the public cloud to form a suspicious missing child. Double-Blinded Finder receives a search query, runs blind face matching, and returns the results to PMC. The PMC browses the first k decrypted photos and sends the confirmation feedback to the cloud. If it matches, the system will notify the CME.

3 Blind Face Matching

3.1 Security Model

Adversaries are anyone who can access the Internet. The attacker can obtain an encrypted photo of the child that can be missing and the private key of the child that is truly missing and can perform a face matching algorithm. Exposure of photos may result in privacy risks for both parties. (1) Before the suspicious missing child, the exposed photos may greatly hurt their privacy, because the photos are usually concentrated on children with dirty clothes, (2) near the real missing children, exposure may lead to minor damage to them. Therefore, the security mechanism should consider the protection of both parties, for example, using a double-blind approach.

In order to achieve blind matching in the cloud under the adversary model, our mechanism should preserve the privacy of suspicious and truly missing children. In addition to matching results, an attacker should not infer any information about a suspicious or real missing child from an encrypted photo and private key. In other words, an attacker should not be able to link two encrypted photos, even if they contain the same child and are uploaded by the same user. In order to meet this requirement, we use the IPE method for blind face matching.

3.2 Definition of IPE

IPE is one specified functional encryption [10] where each decryption key corresponds to a specified function. When the holder of a decryption key for the function f gets an encryption of a message **m**, the only thing his key allows him to learn is

$f(\mathbf{m})$. In IPE method [2], the two parameters of inner product function are expressed as predicate vectors $\boldsymbol{v}^Y \in \Re^n$ (for a secret key, such as the descriptor of the true missing child) and attribute vector $\boldsymbol{v}^D \in \Re^n$ (for a ciphertext, such as the descriptor of the suspicious missing child), where

$$f(\mathbf{m}) = \mathbf{m}, \text{ if } \boldsymbol{v}^D \cdot \boldsymbol{v}^Y = 0 \tag{3}$$

The attribute-hiding IPE scheme [9] consists of a tuple of probabilistic polynomial-time algorithms:

$$\mathcal{A} = \{Setup, KeyGen, Enc, Dec\} \tag{4}$$

- $Setup(1^\lambda, n) \rightarrow (pk, sk)$: Input security parameters 1^λ, set the algorithm output (primary) public key pk and (master) secret key sk.
- $KeyGen(pk, sk, \boldsymbol{v}^Y) \rightarrow sk_T$: When entering the master public key pk, secret key sk, and predicate vector $\boldsymbol{v}^D$, the key generation algorithm outputs a corresponding secret key sk_T.
- $Enc(pk, \mathbf{m}, \boldsymbol{v}^D) \rightarrow C_D$: On input the master public key pk, plaintext $\mathbf{m}$ in some related messages in plaintext space MSG and attribute vector $\boldsymbol{v}^D$, encryption algorithm output ciphertext C_D.
- $Dec(pk, sk_T, C_D) \rightarrow \mathbf{m}$: On input the master public key pk, secret key sk_T and ciphertext C_D. The decryption algorithm outputs either $\mathbf{m}$ or the distinguished symbol $\perp$.

We define: when A as a random variable or distribution, $y \stackrel{R}{\leftarrow} A$ denotes that y is randomly selected from A according to its distribution, then the correctness attribute that the IPE scheme should have is as follows:

- for all $(pk, sk) \stackrel{R}{\leftarrow} Setup(1^\lambda, n)$, all $\boldsymbol{v}^D$ and $\boldsymbol{v}^Y$, all $sk_T \stackrel{R}{\leftarrow} KeyGen(pk, sk, \boldsymbol{v}^Y)$, all message $\mathbf{m}$, all ciphertext $C_D \stackrel{R}{\leftarrow} Enc(pk, \mathbf{m}, \boldsymbol{v}^D)$, it holds that $\mathbf{m} = Dec(pk, sk_T, C_D)$ if $\boldsymbol{v}^D \cdot vecv$=0. Otherwise, it holds with negligible probability.

3.3 Blind Face Matching via IPE

According to the IPE defined above, our blind matching scheme runs on the elliptic curve and proves:

Theorem 1 *The proposed IPE scheme adaptively attribute-hiding for the selected plaintext attack under the decision linearity (DLIN) assumption.*

Attribute hiding means giving a ciphertext $C_{D_b} = Enc(pk,\ textbfm_b,\ vecv_b^D)$ nobody can distinguish it is encrypted with $textbfmmessage_0$ or $textbfm_1$, under the vector $vecv_0^D$ or $vecv_1^D$. Therefore, the privacy of vectors and messages is protected. With this feature, we use two IPE parameters in the Double-Blinded

Finder system to match suspicious missing children and encrypt/decrypt photos separately in the following six main steps.

(1) Setup CME run *Setup* algorithm twice to get the public parameters $PK = (pk_F, pk_I)$ and the master key $SK = (sk_F, sk_I)$, where F and I mean flag and face image, respectively. Then, CME publish the public parameters to the cloud and keep the master key secret.

(2) Request Private Key Pair from PMC The PMC runs the face representation module to get the real missing child's 128d identity vector $\boldsymbol{v}^Y$ and adds the basic fact attribute, which is then submitted to the CME via a secure channel. After receiving the descriptor and reviewing it by the CME, the CME first generates the vector $vecV^Y = (vecV^Y, -1)$, and then runs the key generation algorithm to obtain the matching private key. Decryption private key with $sk_{T,F} = KeyGen(pk_F, sk_F, \; vecV^Y)$ and (missing) sub-face image I with $sk_{T,I} = KeyGen(pk_I, sk_I, \; VECv^Y)$. The PMC downloads the private key pair through the secure channel and then sends the matching private key $sk_{T,F}$ searches for the missing child from the suspicious photos in the cloud, while retaining the decrypted private key $sk_{T,I}$ local at Local client.

(3) Photo Uploading from SNH SNH runs the face representation module to obtain the identity vector $\boldsymbol{v}^D$ of the suspected missing child and generates a vector set $\{\boldsymbol{v}_i^D\}_{i=t_s}^{tS}$, where $\boldsymbol{v}_i^D = (\boldsymbol{v}^D, i)$, the two positive integers t_s and tS are the minimum and maximum identity similarity scores, respectively. Then, SNH encrypts the (suspicious) face I flag F into the password flag C_F and encrypts I as cipher image C_I

$$\begin{aligned} C_F &= \{C_{F,i}\}_{i=t_s}^{tS} \\ C_I &= \{C_F, C_{I,aes}\} \\ C_{F,i} &= Enc(pk_F, \text{“}\mathbf{smc}\text{”}, \boldsymbol{v}_i^D) \\ C_{I,i} &= Enc(pk_I, key_{aes}, \boldsymbol{v}_i^D) \\ C_{I,aes} &= AesEnc(key_{aes}, I) \end{aligned} \tag{5}$$

Where "**smc**" means "suspicious missing child," key_{aes} is a random AES key, $AesEnc$ is AES encryption algorithm [11]. Finally, SNH uploads $\{C_F, C_I\}$ along with the metadata to the cloud, such as where and when the photo was taken. These metadata will help recover the missing child.

(4) Face Matching Once the cloud receives a matching request from PMC with the matching private key $sk_{T,F}$, it orderly extracts the encrypted suspicious faces $\{C_F, C_I\}$ and generates a matching list $\mathcal{L}$ with Alg. 1. Then, the cloud sorts the list by the matching score sc and returns the first k results $\mathcal{L}_k$ to PMC. Since the cipher flag C_F is encrypted by $\{\boldsymbol{v}_i^D\} = \{(\boldsymbol{v}^D, i)\}$ and the matching private key refers to $\boldsymbol{v}^Y = (\boldsymbol{v}^Y, -1)$, the decryption success means that $\boldsymbol{v}_{sc}^D \cdot \boldsymbol{v}^Y = 0$ and thus $\boldsymbol{v}^D \cdot \boldsymbol{v}^Y = sc$.

(5) Confirmation Once the matching list $\mathcal{L}_k$ is received, PMC orderly decrypt the suspicious photos to find their child image list $\mathcal{I}$ with Alg. 1. With those face images in $\mathcal{I}$ and the associated metadata, PMC confirm the matching result and inform CME to seek their child.

(6) Supervision In order to supervise all the photos, CME generate a vector $\boldsymbol{v}^{Y_0} = (255, \cdots, 255)$ and private key pair $\{sk_{T_0,F}, sk_{T_0,I}\}$. It is easy to find that $\boldsymbol{v}^{Y_0}$ could match all the suspicious photos and decrypt them.

3.4 Privacy Analysis

The suspicious photo is encrypted by AES whose key is encrypted by IPE. The feature vector of the suspicious missing child is used to encrypt the flag **"ms"** ("matching success") and the AES key by IPE. From Theorem 1, the privacies of the feature vector and the AES key are protected, thus the privacy of suspicious photos is preserved.

The decryption private key is holded by PMC, thus it does not weaken the security of the suspicious faces. Even the adversary gets the matching private key, he only can determine which child looks like the missing child but cannot recover the feature vector or the missing child face. Thus, the privacy of the true missing child is protected.

4 System Evaluation

In this section, we evaluate Double-Blinded Finder to demonstrate its **effective** and **efficient** performance in finding missing children while protecting **privacy**.

4.1 Experiment on Face Recognition

In order to evaluate the face recognition found by **effective** missing children, we first check the face verification of the LFW test data set [8], then view the face match and search the test set of the LCFW data set. Our face matching method uses a 128d32-bit fixed-point feature vector to represent faces and simple internal products to compare a pair of facial features. In this way, our MT-FaceNet model achieves 92.5We further look at face recognition performance on the test set of LCFW dataset to simulate the process of *Finding Missing Children* as follows:

- **PMC**: 330 child images with 100 identifies (for simulating the true missing children) are randomly picked from dataset. Then, face detection and representation are applied to generate the 128d "*true*" face vectors $\mathcal{V}^Y = \{\boldsymbol{v}_i^Y\}_{i=1}^{330}$. Note that the gender $\boldsymbol{g}_i^T$ and age a_i^T are given by the groundtruth values.

- **SNH**: The rest including 9,670 child images with 1,000 identities are used to simulate the suspicious missing children. Then, the "*suspicious*" face vectors are denoted as $\mathcal{V}^D = \{\boldsymbol{v}_i^D\}_{i=1}^{9,670}$. Note that the gender $\boldsymbol{g}_i^S$ and age a_i^S are given with MT-FaceNet.
- **Face matching**: For each "*true*" face vector $\boldsymbol{v}_i^Y$ (along with gender, age, and its identity d_i^T) from PMC, we perform face matching and searching in $\mathcal{V}^D$ by using Eq. 2 and receive the matching identity set $D = \{d_i^S\}_{i=1}^k$. If there exists a match (e.g., $d_i^T = d_j^S, 1 \le j \le k$), the matching count adds 1.

Top-k accuracy is used to evaluate performance, which is widely used in ImageNet object recognition [3]. k is the number of recall matches. When k was increased from 1 to 20, MT-FaceNet's accuracy reached 31.2% and 98.5%, showing the improvement of FaceNet. The results validate the effectiveness of our approach and the difficulty of facial recognition tasks for children. Note that the feature vector is low and fixed, which means that the proposed method is suitable for actual blind matching.

4.2 Experiment on Blind Matching

We use the elliptic curve $y^2 = x^3 + 2x + 1$ over $\mathbb{F}_{3167}$ to match blind face matching of missing children. As given in Alg. 1, the computational cost of face blind matching is calculated by the fractional iteration number $m = t_S - t_s + 1$ and *Dec* algorithm, the cost of which depends largely on the feature dimension d. Note that d is 129 (identity vector dimension $n = 128$ plus a matching score index), thus the public parameters and the master key contain maximal two $d \times m$ and two $d \times 2(d + 1)$ matrixes of elliptic points, respectively. The matching private key and decryption private key are both d-dimensional vectors. The cipher flag C_F consists of m d-dimensional point vectors and d-bit strings. The cipher image C_I is consisted of $m \times d$ dimensional points vector and d-bit strings, and an AES-encrypted image $C_{I,aes}$. Thus, the matching will take $m \times d$ Weil pairing computations, while the decryption of image will take d Weil pairing computations and an AES decryption computation. In the experiments, we set $m = 300$.

Table 1 shows the time cost of our IPE algorithm running on a 2.3 GHz Intel Core i7 computer with a single thread. Actually, the encryption algorithm could be paralleled while computing $Enc(pk_F,$ "**smc**"$, \boldsymbol{v}_i^D)$, $Enc(pk_I, Key_{aes}, \boldsymbol{v}_i^D)$, and $Dec(pk_F, sk_{\boldsymbol{v}^Y,F}, C_{F,i})$. Generally, considering multi-thread running, hierarchical matching scheme, and distributed deployment, PMC could receive the result about 15 min after require a search. It shows that our system is practical for secure missing children finding.

Table 1 The time cost of blind face matching

Algorithm step	Times (s)
Request private key pair from PMC	0.14
Photo uploading from SNH	0.18
Face matching (per cipher flag)	39.0

5 Conclusion

In this paper, we proposed a novel system for finding missing children in a double-blinded way. Unlike the face recognition system where the server saves plaintext photos, we use IPE policies to limit access control based on face recognition. With IPE, only the parents whose missing child looks like the child in the photos, and the center could browse the suspicious photos. Due to the use of the above mechanisms, our system is safe to run in the public cloud. We apply multi-task deep learning to train face recognition models for children's face matching. The experimental results reveal that our system can achieve practical performance of blind face matching with negligible privacy leakage of both the suspicious and true missing children sides. Our main job in the future is to continue to improve the accuracy of children's face recognition and prevent inaccuracies in children's faces. In addition, IPE solutions, especially internal products with threshold encryption schemes, should be improved to accommodate more effective face recognition applications in a secure environment.

Acknowledgments We thank all the PCs and reviewers. This work is partially supported by the National Natural Science Foundation of China (Grant Nos. 61772047, 61772513), the Open Project Program of State Key Laboratory of Virtual Reality Technology and Systems, Beihang University (No. VRLAB2019C03), the Open Funds of CETC Big Data Research Institute Co.,Ltd., (Grant No. W-2018022), the Science and Technology Project of the State Archives Administrator (Grant No. 2015-B-10), and the Fundamental Research Funds for the Central Universities (Grant Nos. 328201803, 328201801).

References

1. B. Amos, L. Bartosz, M. Satyanarayanan, Openface: A general-purpose face recognition library with mobile applications. Tech. rep., CMU-CS-16-118, CMU School of Computer Science (2016)
2. J. Katz, A. Sahai, B. Waters, Predicate encryption supporting disjunctions, polynomial equations and inner products, in *EUROCRYPT*, vol. 17, pp. 146–162 (2008)
3. A. Krizhevsky, I. Sutskever, G. Hinton, Imagenet classification with deep convolutional neural networks, in *NIPS*, pp. 453–464 (2012)
4. H. Lu, Y. Li, C. Min, H. Kim, S. Serikawa, Brain intelligence: Go beyond artificial intelligence. Mobile Networks Appl. **23**(7553), 368–375 (2017)
5. H. Lu, Y. Li, S. Mu, D. Wang, H. Kim, S. Serikawa, Motor anomaly detection for unmanned aerial vehicles using reinforcement learning. IEEE Internet Things J. **5**(4), 2315–2322 (2018). https://doi.org/10.1109/JIOT.2017.2737479, https://doi.org/10.1109/JIOT.2017.2737479

6. H. Lu, Y. Li, T. Uemura, H. Kim, S. Serikawa, Low illumination underwater light field images reconstruction using deep convolutional neural networks. Futur. Gener. Comput. Syst. **82** (2018)
7. H. Lu, D. Wang, Y. Li, J. Li, X. Li, H. Kim, S. Serikawa, I. Humar, Conet: A cognitive ocean network. CoRR **abs/1901.06253** (2019). http://arxiv.org/abs/1901.06253
8. E.L. Miller, G.B. Huang, A. RoyChowdhury, et al., Labeled faces in the wild: A survey. *Advances in Face Detection and Facial Image Analysis*, pp. 189–248 (2016)
9. T. Okamoto, K. Takashima, Adaptively attribute-hiding (hierarchical) inner product encryption. IEICE Trans. Fundam. Electron. Commun. Comput. Sci. **E99-A**(1), 92–117 (2016)
10. A. Sahai, B. Waters, Functional encryption: beyond public key cryptography. Tech. rep., INS (2008)
11. F. Schroff, D. Kalenichenko, J. Philbin, Facenet: A unified embedding for face recognition and clustering, in *IEEE CVPR*, pp. 815–823 (2015)
12. S. Serikawa, H. Lu, Underwater image dehazing using joint trilateral filter. Comput. Electr. Eng. **40**(1), 41–50 (2014). https://doi.org/10.1016/j.compeleceng.2013.10.016, https://doi.org/10.1016/j.compeleceng.2013.10.016

Complex Object Illumination Transfer Through Semantic and Material Parsing and Composition

Xiaodong Li, Rui Han, Ning Ning, Xiaokun Zhang, and Xin Jin

1 Introduction

Object relighting without 3D models in the field of visual computing has been widely studied [1–3].

Object relighting can be divided into two categories [4–7]: simple object image relighting and complex object image relighting. In addition to the 3D reconstruction theory, another research route is applied practically example image to the real world [8–16].

In the simple object relighting, it is easy to find a reference object image similar to the input image geometry in the reference object image library. Therefore, the method of relighting a simple object image usually uses two reference object images. However, it is difficult to find a complex object image to find a reference object image similar to its geometric shape in the reference object image library, and the above method is no longer applicable. In this paper, the inspiration of object semantic analysis and synthesis method is obtained. Semantic segmentation of the

---

X. Li
Beijing Electronic Science and Technology Institute, Beijing, China

CETC Big Data Research Institute Co., Ltd., Guizhou, China

R. Han · N. Ning · X. Zhang
Beijing Electronic Science and Technology Institute, Beijing, China

X. Jin (✉)
Beijing Electronic Science and Technology Institute, Beijing, China

State Key Laboratory of Virtual Reality Technology and Systems, Beihang University, Beijing, China
e-mail: jinxin@besti.edu.cn

Y. Li, H. Lu (eds.), *3rd EAI International Conference on Robotic Sensor Networks*, EAI/Springer Innovations in Communication and Computing,
https://doi.org/10.1007/978-3-030-46032-7_5

image of the complex input object into each component is used to relight, and then the components of the relighting are semantically analyzed to obtain the relighting result.

Unlike simple objects have the same material, complex objects are made up of different materials. Selecting the reference object image of different materials has a great influence on the illumination migration of the input object image. In this paper, the material analysis of the object is used to analyze the materials of the components that make up the complex object, and then the reference of the same material is found in the reference object image library. The images after material analysis can improve the accuracy of illumination transfer.

2 Related Work

In 2016, Jin and Tian et al. proposed a simple object illumination transfer method. They need a pair of example object. The example object A has the same illumination condition as the subject object, while the example object B has the example illumination. They firstly changed the example object's color according to the subject object. By image patch match, the modified example object A was warped to the subject object. The modified example object B was warped to the subject object in the same way. At last, they used transfer model from the warped example object A and B to transfer the illumination of the example object B to the subject object [17].

In 2015, Wang et al. proposed a joint solution for solving semantic objects and partial segmentation, providing a higher object-level context to guide partial segmentation, more detailed partial-level positioning for refining object segmentation. They introduced the concept of the SCP (semantic component part), that is, similar semantic parts between different objects were grouped and shared. A dual stream FCN (Complete Convolution Network) was trained aim to provide SCP and object potential at each of pixel. Meanwhile, a compact set of segments was obtained from the SCP network prediction. Given the potential and generated segments, they ultimately constructed a valid FCRF to predict the final object and part of the tag to explore the remote context [18].

In 2015, Bell et al. introduced a large-scale, open wild material dataset, it is named MINC (Material in Context Database), and then implemented wild image material identification and segmentation by combining this dataset with deep learning. The MINC is an order of magnitude. It is larger than the previous material database, with 23 categories being more diverse and well sampled. They used MINC to train CNN (Convolutional Neural Network) to perform two tasks: classify materials from patches, identify and segment the material of the complete image [19].

In 2005, Geusebroek et al. introduced 1,000 objects collected by ALOI recorded in various imaging environments. In order to capture sensory changes in object records, they systematically changed the illumination angle of each object, illuminated the color and angle of view, and additionally captured a wide baseline stereo image. They recorded each object on more than hundred images and collected a total of 110,250 image collections that could be made publicly available for scientific purposes [20].

3 Complex Object Illumination Transfer

3.1 Method Overview

The method can be divided into four steps. As shown in Fig. 1: First, the complex object image composed of different materials was segmented into several object component images by object semantic analysis, and then the reference material image matching the input object component image material was selected in the reference object image library by using the object material analysis. For each topic object part, we need two sample objects, one with the same illumination as the theme object part and the other with example illumination. Secondly, the selected reference object was subjected to block matching deformation to obtain a deformed reference image. Then transform the deformed reference object image and the input object image component through the local and global affine transformation model, and migrate the illumination state of the reference object image to the input object image component to obtain the input object image component in the reference object image. The result of the relighting of the input object image component after the migration of the reference object image illumination condition was obtained. Finally, the image components of the input objects after the relighting were synthesized into the image results of the relighting input object through object semantic analysis.

3.2 Semantic and Material Parsing and Composition

As shown in Fig. 2, we firstly determined the overall semantics of the complex input image, and then analyzed the semantic tags of the components of the input image. The semantic label of each component of the input image was determined, and the complex input image was segmented to obtain the component image of each part of the input image. Then we matched the corresponding sample object, we can see that the matching object A is an inappropriate match, and the matching object B is an appropriate match, which is caused by material analysis.

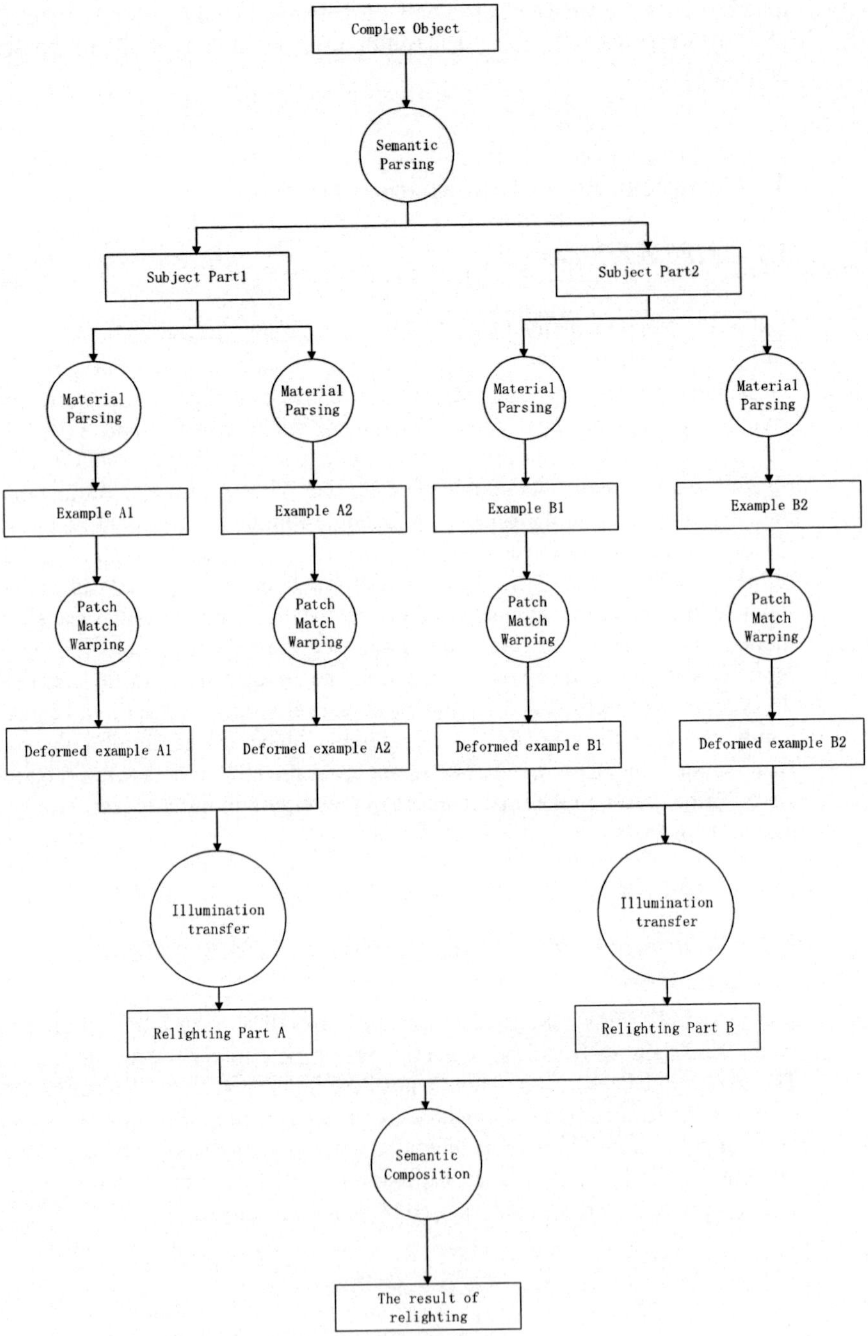

Fig. 1 The process of complex object illumination transfer

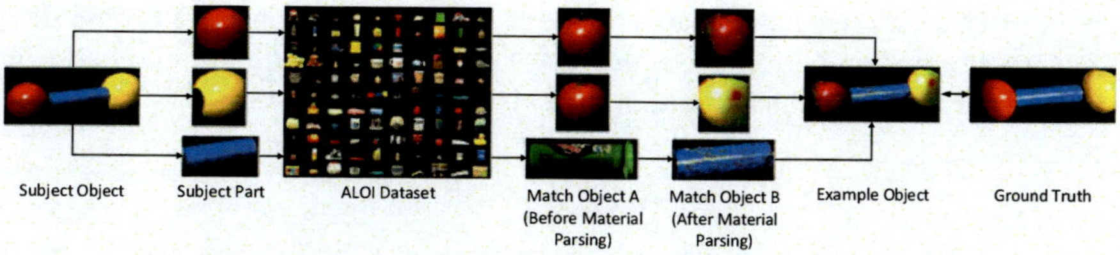

Fig. 2 Object match comparison (before material parsing and after material parsing)

3.3 Patch Match Warping

A general object has a different shape even if it has the same semantic tag. We tried to find a pair of corresponding pixels E from the example object to the subject object S. This high-density matching consumes a lot of time and space. Thus, for each patch of subject object S, we seek a patch of example object E that looks similar. L_2 norm is used for square patches, whose side length is $2r + 1$. For pixels $p(p \in s)$and the relevant pixel $q(q \in \varepsilon)$, we minimize the following formula:

$$\sum_{i=-r}^{r} \sum_{j=-r}^{r} \|S(x+i, y+j) + E(x+i, y+j)\|^2 \quad (1)$$

where (x, y) is the coordinate pixel in the patch. We solved the warp by the generalized patch match correspondence algorithm [21].

Through the above patch match warping method, the example object E was warped according to the subject object S.

3.4 The Local and Global Transfer

We proposed a transfer model based local and global transfer for illumination transfer, which is learned from the warped example object A and the warped example object B and is applied to the subject object S ,whose output is the relighting subject object result (denoted as R).

We used $P_{k(\bullet)}$ to denote the kth patch of the image. For each patch which containing N pixels, $P_{k(\bullet)}$ is a $1 \times N$ vector. The local transformation is represented as T_k for each patch.

We use the same transformation T_k to convert the subject object S to the output relighting subject object result R.

Then, we need a regularization term to avoid overfitting. We choose the global transformation G from the entire image of A to B. Thus, the whole energy function of our local and global transfer model is defined as follows:

$$R = \arg\min_{R,\{T_k\}} \sum_k \| P_k(R) - T_k(I) \|^2 + n \sum_k \| P_k(B) - T_k(A) \|^2 + m \sum_k \| T_k - G \|^2 \tag{2}$$

where n and m are the relative importance of each term. In our method, we made $n = 0.01, m = 1$ (pixel value 2 [0; 255]), $N = 25(5 \times 5$ patch). The above minimization could be solved as a standard local liner regression [22].

4 Experimental Results

The experimental materials of this experiment are all derived from the ALOI dataset. As shown in Fig. 3, the example object A has the same illumination condition as subject object, while the example object B has different illumination condition. Compared with the ground truth, the relighting subject object could perform the same illumination condition as the example object B.

In first way, we tested our method on one subject object with two example objects, as shown in Fig. 4. To obtain novel illumination effects, we used different example objects with different semantic label and geometry and same material to relight the subject object.

In second way, we tested our method on one subject object with multiple example objects. As shown in Fig. 5, to obtain multiple illumination effects, we could use sample objects with similar semantic labels , the same material labels, and different geometries. The relighting subject object by different geometry example objects is similar although slightly different in shadow, for which is caused by the different normal of different geometry example objects. To be truth, we could convincingly generate multiple relighting results on the subject object under multiple example objects.

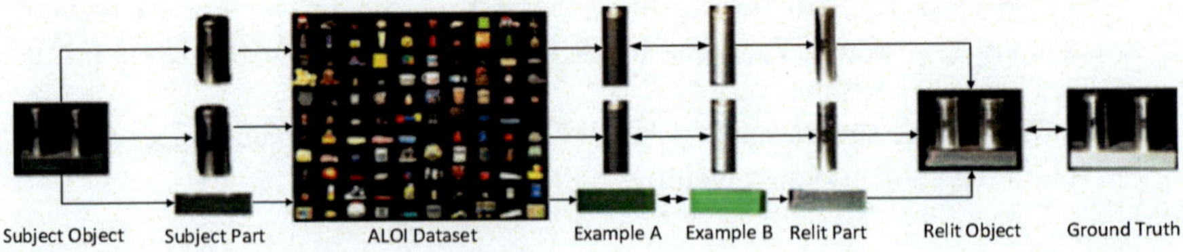

Fig. 3 An experiment result example of complex object illumination transfer

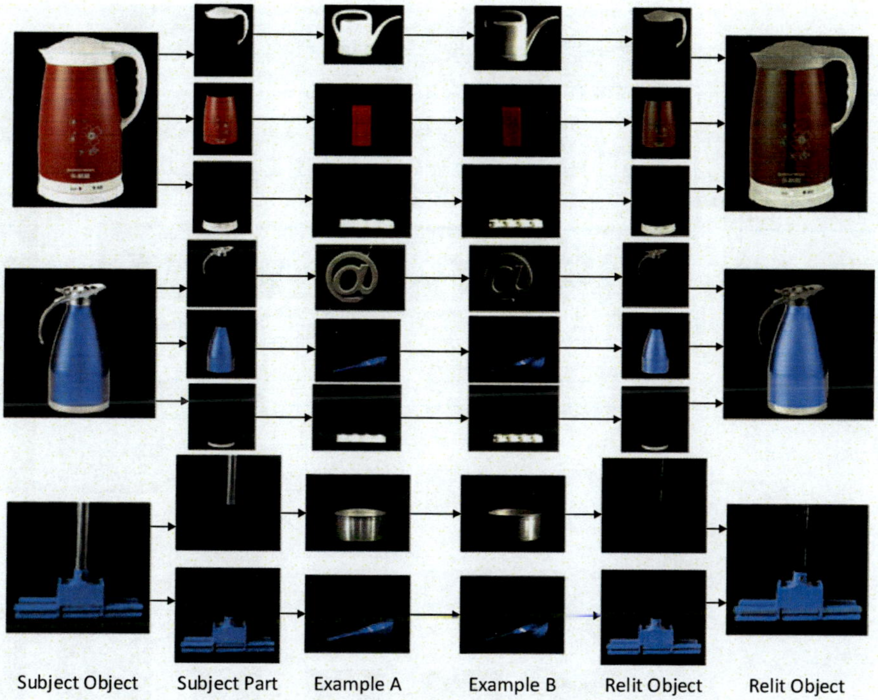

Fig. 4 Experiment result examples of one subject object with two example objects

5 Conclusion

In our work, we proposed a novel complex object illumination transfer based semantic and material parsing and composition method. In this method, object semantic and material parsing and composition are used to process complex objects into simple objects, reducing the difficulty of complex object illumination transfer methods. Patch match warping is used to perform block level matching on the object foreground area and the input object image in the reference object image, find similar blocks corresponding to the input object image, and rearrange the reference object image at the block level. Local and global transfer for illumination is used to effect the migration of reference object image illumination effects into the input object image. Experiments show that the complex object illumination transfer based semantic and material parsing and composition method can be applied to practical complex object illumination transfer applications.

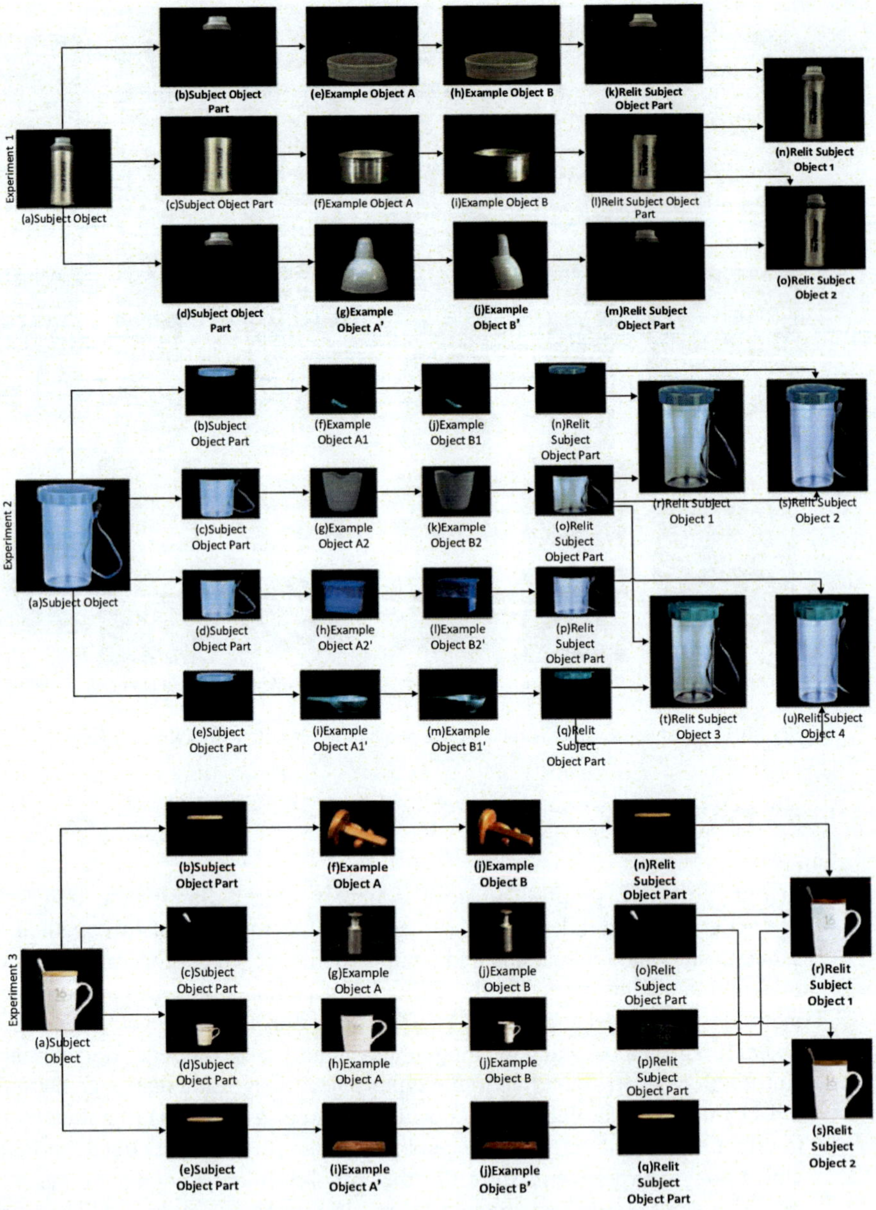

Fig. 5 Experiment result examples of one subject object with multiple example objects

Acknowledgments We thank all the PCs and reviewers. This work is partially supported by the National Natural Science Foundation of China (Grant Nos. 61772047, 61772513), the Open Project Program of State Key Laboratory of Virtual Reality Technology and Systems, Beihang University (No. VRLAB2019C03), the Open Funds of CETC Big Data Research Institute Co.,Ltd., (Grant No. W-2018022), the Science and Technology Project of the State Archives Administrator (Grant No. 2015-B-10), and the Fundamental Research Funds for the Central Universities (Grant Nos. 328201907).

References

1. X. Jin, Y. Li, N. Liu, X. Li, X. Jiang, Y. Tian, S. Ge, Scene relighting using a single reference image through material constrained layer decomposition, in *International Symposium on Artificial Intelligence and Robotics , Kitakyushu, Japan, 25–26 November, ISAIR 2017* (Springer, 2017), pp. 37–44
2. X. Jin, Y. Li, N. Liu, X. Li, X. Jiang, C. Xiao, S. Ge, Single reference image based scene relighting via material guided filtering. Opt. Laser Technol. **110**, 7–12 (2019). Special Issue: Optical Imaging for Extreme Environment
3. X. Chen, X. Jin, K. Wang, Lighting virtual objects in a single image via coarse scene understanding. SCIENCE CHINA Inf. Sci. **57**(9), 1–14 (2014)
4. X. Chen, X. Jin, H. Wu, Q. Zhao, Learning templates for artistic portrait lighting analysis. IEEE Trans. Image Process. **24**(2), 608–618 (2015)
5. X. Chen, X. Jin, Q. Zhao, H. Wu, Artistic illumination transfer for portraits. Comput. Graph. Forum **31**(4), 1425–1434 (2012)
6. X. Chen, K. Wang, X. Jin, Single image based illumination estimation for lighting virtual object in real scene, in *12th International Conference on Computer-Aided Design and Computer Graphics, CAD/Graphics 2011, Jinan, China, September 15–17*, pp. 450–455, 2011
7. X. Jin, M. Zhao, X. Chen, Q. Zhao, S.-C. Zhu, Learning artistic lighting template from portrait photographs, in *Computer Vision - ECCV 2010, 11th European Conference on Computer Vision, Heraklion, Crete, Greece, September 5–11, 2010, Proceedings, Part IV*, pp. 101–114, 2010
8. Z. Quan, W. Yang, G. Gao, W. Ou, H. Lu, C. Jie, L.J. Latecki, Multi-scale deep context convolutional neural networks for semantic segmentation. World Wide Web-Internet Web Inf. Syst. **22**(7), 1–16 (2018)
9. Q. Zhou, J. Cheng, H. Lu, Y. Fan, S. Zhang, X. Wu, B. Zheng, W. Ou, L. Jan Latecki, Learning adaptive contrast combinations for visual saliency detection. Multimed. Tools Appl. 1–29 (2018)
10. Z. Quan, Z. Cheng, W. Yu, Y. Fan, Z. Hu, X. Wu, W. Ou, W. Zhu, L. J. Latecki, Face recognition via fast dense correspondence. Multimed. Tools Appl. **12**, 1–19 (2017)
11. Z. Quan, B. Zheng, W. Zhu, L.J. Latecki, Multi-scale context for scene labeling via flexible segmentation graph. Pattern Recogn. **59**(C), S0031320316300085 (2016)
12. S. Serikawa, H. Lu, Underwater image dehazing using joint trilateral filter. Comput. Electr. Eng. **40**(1), 41–50 (2014). [Online]. Available: https://doi.org/10.1016/j.compeleceng.2013.10.016
13. H. Lu, Y. Li, S. Mu, D. Wang, H. Kim, S. Serikawa, Motor anomaly detection for unmanned aerial vehicles using reinforcement learning. IEEE Internet Things J. **5**(4), 2315–2322 (2018). [Online]. Available: https://doi.org/10.1109/JIOT.2017.2737479
14. H. Lu, Y. Li, C. Min, H. Kim, S. Serikawa, Brain intelligence: Go beyond artificial intelligence. Mobile Networks Appl. **23**(7553), 368–375 (2017)
15. H. Lu, D. Wang, Y. Li, J. Li, I. Humar, Conet: A cognitive ocean network. CoRR **abs/1901.06253** (2019). http://arxiv.org/abs/1901.06253

16. H. Lu, Y. Li, T. Uemura, H. Kim, S. Serikawa, Low illumination underwater light field images reconstruction using deep convolutional neural networks. Futur. Gener. Comput. Syst. **82** (2018)
17. X. Jin, Y. Tian, N. Liu, C. Ye, J. Chi, X. Li, G. Zhao, Object image relighting through patch match warping and color transfer. IEEE **09**, 235–241 (2016)
18. P. Wang, X. Shen, Z. L. Lin, S. Cohen, B.L. Price, A.L. Yuille, Joint object and part segmentation using deep learned potentials, in *2015 IEEE International Conference on Computer Vision, ICCV 2015, Santiago, Chile, December 7–13*, pp. 1573–1581, 2015
19. S. Bell, P. Upchurch, N. Snavely, K. Bala, Material recognition in the wild with the materials in context database, in *IEEE Conference on Computer Vision and Pattern Recognition, CVPR 2015, Boston, MA, USA, June 7–12*, pp. 3479–3487, 2015
20. J. Geusebroek, G.J. Burghouts, A.W.M. Smeulders, The amsterdam library of object images. Int. J. Comput. Vis. **61**(1), 103–112 (2005)
21. C. Barnes, E. Shechtman, D.B. Goldman, A. Finkelstein, The generalized patchmatch correspondence algorithm, in *Computer Vision - ECCV 2010, 11th European Conference on Computer Vision, Heraklion, Crete, Greece, September 5–11, Proceedings, Part III*, pp. 29–43, 2010
22. Y. Shih, S. Paris, F. Durand, W.T. Freeman, Data-driven hallucination of different times of day from a single outdoor photo. ACM Trans. Graph. **32**(6), 200:1–200:11 (2013)

Global-Best Leading Artificial Bee Colony Algorithms

Di Zhang and Hao Gao

1 Introduction

Artificial bee colony (ABC) algorithm is one of the competitive evolutionary algorithms (EAs) first proposed by Karaboga [1] in 2005 by simulating the foraging behavior of the honeybee swarm. Compared with other population-based optimization methods, it shows similar or even superior search performance for many hard optimization problems, especially in multimodal and multidimensional problems due to its strong global search ability. Regarding its few controlling parameters and easy implementation property, ABC has been successfully applied to many practical optimization problems, such as antenna design [2, 3], magnetics [4, 5], neural network controlling [6, 7], and system engineering [8–13].

ABC has been demonstrated to have strong global search ability, observations, however, it still faces a challenge of slow convergence speed when compared with PSO and DE especially in simple problems [14]. The reasons can be described as follows. Both exploration and exploitation abilities are important factors during the optimization process. Exploration refers to the ability of jumping out of the local optima, and exploitation refers to applying the previous knowledge of previous solutions' information to fast converge to a potential global optimum. However, the two aspects contradict with each other, and this issue should be handled to achieve better performance. Therefore, to achieve the goals, plenty of works have been studied [15–19]. Inspired by PSO, Zhu and Kwong [15] introduce the global best information to ABC to enhance its exploitation ability and accelerate the convergence speed. To model the behavior of foragers more accurately, Karaboga and Gorkemli [16] raise a new definition, the best potential solution among the

D. Zhang · H. Gao (✉)
College of Automation and College of Artificial Intelligence, Nanjing University of Posts and Telecommunications, Nanjing, China

Y. Li, H. Lu (eds.), *3rd EAI International Conference on Robotic Sensor Networks*, EAI/Springer Innovations in Communication and Computing,
https://doi.org/10.1007/978-3-030-46032-7_6

neighbors of a selected bee, to make a better local search in the onlooker bees phase. Gao and Liu [17] present a new search equation influenced by DE; it guides the bees to only search on the current best solution to accelerate the convergence speed. Furthermore, to make a better balance between exploration and exploitation of ABC, they proposed a new equation by utilizing two random selected food sources and incorporating an orthogonal learning method to enhance its performance [18]. Akay and Karaboga [19] make a fine tuning of parameters in the standard ABC's search equation. They propose that two perturbations influence the performance mostly: frequency and magnitude, which control how many dimensions to update and how long the step should be made, respectively. Thus, the frequency perturbation is related to convergence speed, and magnitude perturbation corresponds to the ability of avoiding getting stuck in local minima.

Regarding the inefficiency search strategy of the standard ABC, this chapter proposes two global-best leading methods, GLABC-pso and GLABC-de, to improve the performance of the original algorithm. In particular, GLABC-pso inherits the advantages of PSO [20], and GLABC-DE utilizes the character of DE [21] to get much more efficient search abilities in the employed bee phase and accelerate its convergence speed. Besides, they both apply a global-best leading strategy to further enhance their achievements. To evaluate the performance of our proposed algorithm on both the benchmark and real-world problems, a set of benchmark functions and the large-scale transmission pricing problem are employed in this chapter. The rest of the chapter is organized as follows.

Section 2 presents the standard artificial bee colony and gives a further analysis of its performance. The proposed two GLABCs are described in Sect. 3 and in Sect. 4; experiments on a set of benchmark functions are conducted to evaluate the performance of the proposed algorithm. Section "Acknowledgments" concludes this chapter.

2 ABC Framework

Inspired by the intelligence of foraging behavior in the honey colony, ABC is a population-based algorithm first proposed by Karaboga [1] in 2005. It consists of three groups of bees: employed bees, onlooker bees, and scout bees. Employed bees work to forage food sources which are visited by themselves previously and only search around their vicinity. When they get nectar, employed bees return to the hive and play a waggle dance to propagate the information of their food sources to the other colonies in the hives. As a response, onlooker bees choose one or a number of employed bees to follow and make exploitation search around the food sources. A bee regenerated in the search space for enhancing the global search ability of ABC is called scout bees. When the food source found by an employed bee has not been improved for a long time, the corresponding employed bee will become a scout and starts random search across the search space until it finds another good source and becomes employed again. The scout bees' work does not depend on the employed or onlooker bees' finding; it explores new areas to get new food source and reveals this information to the hive.

For every food source, there are only one employed bee and a number of onlooker bees. In the whole population, half of them are employed bees, and the others are onlooker bees. The food sources' number is equal to the number of employed bees. To make further explanation of artificial bee colony, we divided it into five units:

2.1 *Initialization Process*

Considering the search space is the environment of food sources, ABC begins with randomly initialized food source sites related to the solutions in the search space. Let *NP* denote the number of food sources and *D* represent the dimension of the search space; the initialize formulation can be represented as

$$x_{ij} = x_j^{\min} + \text{rand}\,(0, 1)\left(x_j^{\max} - x_j^{\min}\right) \tag{1}$$

where $i = 1, 2, \ldots, NP$, $j = 1, 2, \ldots, D$. $x_j^{\max}$ and $x_j^{\min}$ are the upper and lower bounds for *j*th variable in the solution space, respectively.

Sending Employed Bees to the Food Source

As aforementioned, employed bees are responsible for exploration of food sources depending on a memory of their previous position and a randomly selected neighbor food source. The search equation can be defined as

$$v_{ij} = x_{ij} + \phi_{ij} * \left(x_{kj} - x_{ij}\right) \tag{2}$$

where x_k represents a neighbor of food source x_i, which is randomly selected from the current population, $k \neq i$, $\phi_{i,j}$ is a uniformly distributed random number in the range $[-1, 1]$, and *j* is a randomly chosen index in $[1, D]$.

After producing the candidate food source v_i, a greedy selection mechanism is applied between x_i and v_i; then the fitness value of a minimization problem can be calculated as

$$\text{Fit}_i = \begin{cases} 1/\left(1 + \text{Obj}_i\right) & \text{if } \text{Obj}_i \geq 0 \\ 1 + \text{abs}\left(\text{Obj}_i\right) & \text{if } \text{Obj}_i < 0 \end{cases} \tag{3}$$

where the Obj_i denotes the objective function value of x_i.

2.2 Calculating Probability Using Roulette Wheel Selection

When all employed bees complete their works, they share the food source information containing the nectar amounts and the positions with onlooker bees in the dance area. Onlooker bees evaluate nectar information and choose a food source to make exploitation around it according to a probability related to its nectar amount. In the standard version of ABC, roulette wheel selection scheme is applied to involve the probability value in which each slice is in proportion to the fitness value, denoted as

$$\mathrm{prob}_i = \mathrm{Fit}_i / \sum_{j=1}^{NP} \mathrm{Fit}_j \tag{4}$$

Obviously, the higher the fitness value, the more probability to be selected.

2.3 Food Source Selection by Onlooker Bees

Once the food source is selected according to the nectar amount, it will be further exploited in the around area. The search equation of onlooker bees is the same with employed bees, as shown in Eq. (2). Then a greedy selection is applied between x_i and v_i similar to the employed bee phase.

2.4 Scout Production

If a food source has not been improved after a predetermined number of cycles, it will be abandoned, and the related employed bee will become a scout to investigate a new position with no information of the hive. The scout produces a food source randomly by using Eq. (1).

There are two differences between ABC and other EAs; one is the one-dimension search strategy, and another is their search equation. In this chapter, we mainly focus on studying the difference of search strategy between the ABC with the particle swarm optimization (PSO) and differential evolution (DE). In PSO algorithm, there is a velocity related to each individual, as shown in Eq. (5). In order to facilitate the comparisons on their search abilities, we transform the position update equation of the DE and the ABC into velocity update equation, as shown in Eqs. (6) and (7).

$$vstep_{pso} = w * vstep_{pso} + c1r1\,(gbest - x_i) + c2r2\,(pbest - x_i) \tag{5}$$

where w is the inertia weight, $c1 = c2 = 2$, $r1$ and $r2$ are two random distributed real value in the range [0,1], $g$best is the current global best position found by the population so far, and $p\text{best}_i$ is the personal best position found by individual i so far.

$$v\text{step}_{de} = (x_{r1} - x_i) + F\,(x_{r2} - x_{r3}) \tag{6}$$

where $r1$, $r2$, $r3$ are randomly selected indices in the range $[1, NP]$, where $r1 \neq r2 \neq r3 \neq i$, F is the mutation scale, and $F \in [0, 1]$.

$$v\text{step}_{abc} = \phi\,(x_k - x_i) \tag{7}$$

where $\phi_t \in [-1, 1]$.

The population distribution of the compared algorithms shown in Fig. 1 also proves our investigations. The first row of Fig. 1 is the ABC's population distribution, and it reaches a stable level at 10^{-8} at 300th, 500th, and 800th generations. The distribution of population in the DE decreases rapidly with the increasing of iteration, which gets 10^{-41}, 10^{-59}, and 10^{-111} at 300th, 500th, and 800th generations, respectively. From the third row shown in Fig. 1, the results reveals that the distribution of population in PSO is larger than the ABC at 300th and 500th, and it decreases rapidly to 10^{-13} at 800th generation.

3 Improved ABC Algorithms

As analyzed above, the standard ABC shows slow convergence speed in the optimization process. So it is necessary to propose a new search strategy to get better performance. In this chapter, two global-best leading ABC algorithms (GLABC-pso and GLABC-de) are proposed. Regarding the divided work of bees, it is unreasonable to use the same search equation for all the bees. Hence, in order to strengthen the power of the onlooker bees, we apply a global-best leading strategy into the onlooker bees for searching in a valuable region on the basis of guaranteeing the global search ability. Besides, to accelerate the convergence speed and get more precise results, the property of PSO and DE is introduced into the employed bees, forming two global-best leading artificial bee colony algorithms, denoted as GLABC-pso and GLABC-de, respectively. The detailed description of these two proposed algorithms with a global-best leading strategy is stated as follows.

3.1 GLABC-pso

In this proposed algorithm, we introduce the merits of the particle swarm optimization in the employed bee phase of the ABC. Each employed bee is assigned with a velocity item. Considering the greedy selection scheme in ABC, the current position of each food source is the same with the personal best position found so far in the PSO. Then the third term of the velocity in PSO can be neglected. Besides, as described in Sect. 2, the standard velocity update equation of the PSO shows a global search ability in the beginning but a precise search ability in the end of iteration. For accelerating its convergence speed and making a global search at the beginning of iteration, the velocity equation of GLABC-pso can be formulated as

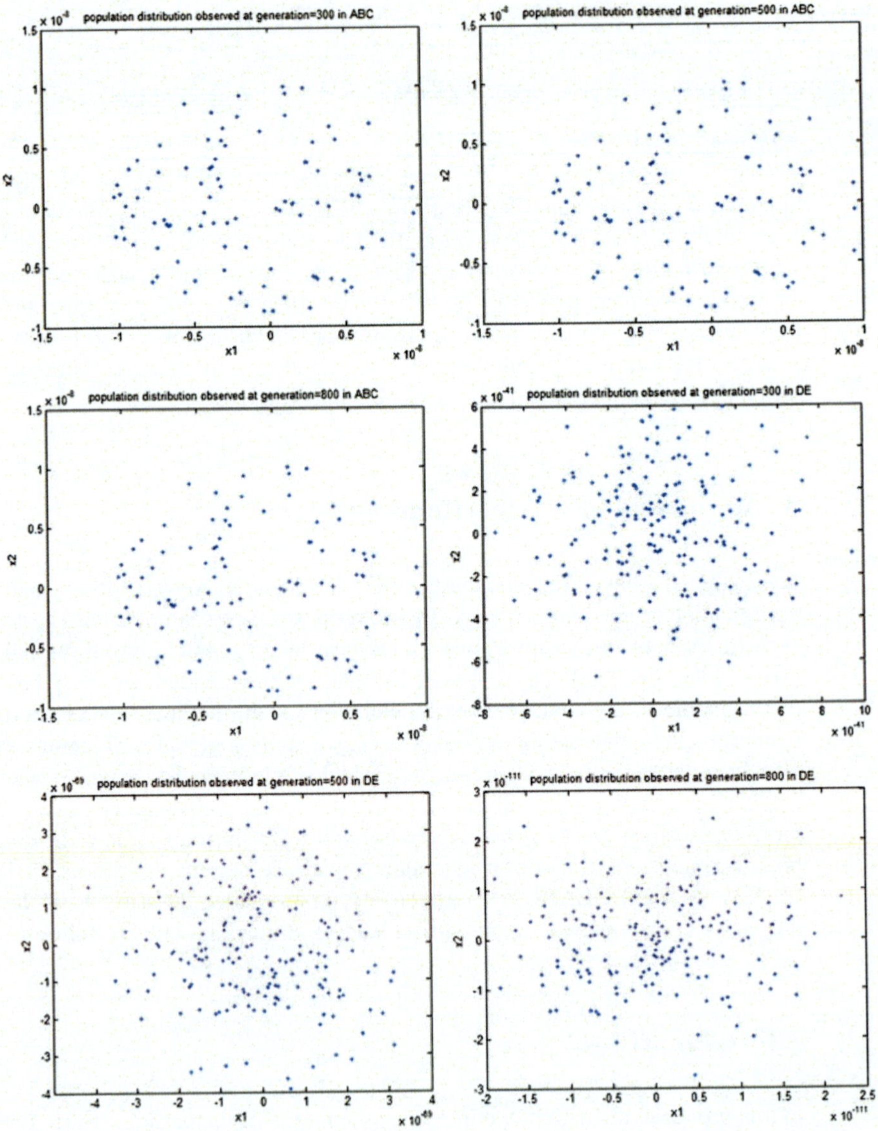

Fig. 1 Population distribution observed at generation = 300, 500, 800 for ABC, DE, and PSO

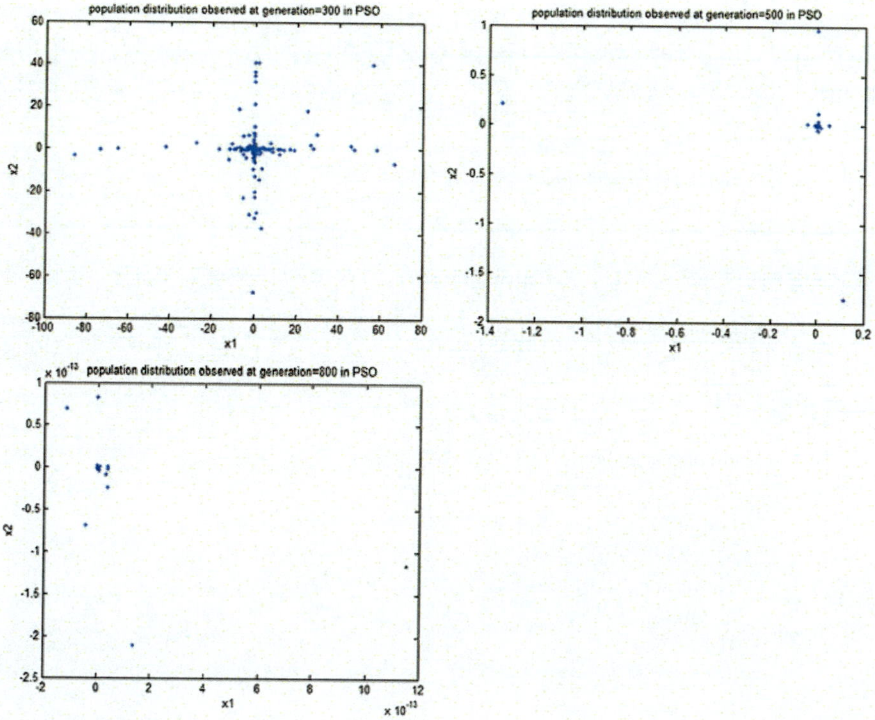

Fig. 1 (continued)

$$vstep_{ij} = c1 * \text{rand} * \left(\text{gbest}_j - x_{ij}\right) \tag{8}$$

where $c1$ is fixed to 2. Then the position update equation can be modified as

$$v_{ij} = x_{ij} + vstep_{ij} = x_{ij} + c1 * \text{rand} * \left(\text{gbest}_j - x_{ij}\right) \tag{9}$$

Algorithm 1: GLABC

Step 1) Initialization

Step 1.1) Randomly generate n individuals in the entire search space

Step 1.2) Evaluate the objective (Obj) and Fitness (Fit) function values of the population

Step 1.3) fes = NP

(continued)

Step 2) The employed bee phase
For $i = 1$ to FoodNumber
Randomly select one dimension: j;
Generate a candidate solution vi using Eq. (5) or (6);
Evaluate Objvi and Fitvi, set fes = fes +1;
If $Fitv > Fit_i$
$x_i = v_i$,
end If
end For
Step 3) Calculate the probability values using Eq. (4);
Step 4) The onlookers phase
Set $i = 1$;
While $t <= NP$
If $rand < prob(i)$
Randomly select one dimension: j;
Randomly select two neighbors of Foods i:
$r1, r2$ $(r1 \neq r2 \neq i)$;
Generate a candidate solution vi using Eq. (13);
Evaluate $Objv_i$ and $Fitv_i, fes = fes + 1$;
If $Fitv > Fit_i$

GLABC-de

In the proposed GLABC-de algorithm, we introduce the search strategy of the DE to the employed bee phase in ABC with a modification on its parameter. The search equation is denoted as

$$v_{ij} = x_{r1j} + F_i \left(x_{r2j} - x_{r3j} \right) \tag{10}$$

where $r1$, $r2$, $r3$ are three random indices selected from the population. F_i is the mutation scale related to each individual. It is independently generated by a Gaussian distribution with a mean value $Fm = 0.5$ and a standard deviation $stf = 0.1$. Compared with a fixed value at 0.5 or other values in the standard DE, a Gaussian generated value can help F_i distribute in a large range, mostly in the interval [0.2, 0.8] and some far beyond of it [22]. Hence, individuals in our algorithm have opportunities to jump out of the local optima.

For the onlooker bee phase, both GLABC-pso and GLABC-de apply a gbest leading search strategy. In order to further enhance their achievements, we use the global best position to guide the onlooker bees' search, denoted as

$$v_{ij} = \text{gbest}_{ij} + M_I * \left(x_{r1j} - x_{r2j} \right) \tag{11}$$

where x_{r1} and x_{r2} are two neighbors of x_i which are randomly selected from the population, $r1 \neq r2 \neq i$. M_i is generated by a Gaussian distribution with the mean value $Mm = 0$ and standard deviation $stm = 0.3$. By using the Gaussian distribution, the values of M_i are mostly around the mean value 0. Thus, it helps the onlooker bees to make more exploitation around the current global best position and accelerate the convergence speed. Besides, the Gaussian distribution also generates relatively large values to enhance the search ability of individuals. Thus, the onlooker bees can be equipped with both exploitation and exploration abilities. The pseudo-code of the two global-best leading algorithms is presented in Algorithm 1.

Experiments and Results

In this section, we conduct an integrated evaluation to estimate the achievements of our algorithms. We present the explanation of testing benchmarks of this paper in the first part of this section. Comparison experiments and results among several state-of-the-art algorithms are conducted and presented in the second part of this section. In the third part, each part of the GLABCs is tested separately to further evaluate their effectiveness.

Benchmark Functions

We employ six benchmark functions [23–28] to demonstrate the effectiveness of our algorithm in Table 1, which describes their formula, initialization range, and global solution in detail. All the functions, in which $f1$–$f3$ are unimodal and $f4$–$f6$ are multimodal functions, respectively, are tested on 10, 20, and 30 dimensions. In this subsection, we first compare the proposed two GLABC algorithms with the standard ABC and eight ABC variants: GABC [15], qABC [16], OCABC [18], ABC (ASF-MR) [19], MABC [14], Gaussian-ABC [25], AABC [26], and STOC-ABC [27]. For a fair comparison, all the algorithms are conducted in 30 independent runs with the same random initialization. The total population size is set to 40, and the maximum iteration number is set to be 1000* dimension, with the maximum number of function evaluations equals NP$_*$MAXITER corresponding. For clarity, the best result of algorithms is marked with boldface front, and the underline indicates that the proposed algorithm achieves similar performance with the best algorithm. The mean and standard deviation of the results achieved by each ABC variant are summarized in Table 2.

Comparison of State-of-the-Art Algorithms

From the results, we notice that both the two proposed GLABCs make great improvement unimodal functions. The reason is that the two update equations inspired by PSO and DE for the employed bees make a faster convergence speed

Table 1 The formula of six benchmark functions

Benchmark functions	Formula	Initialization range	Global optimum
$f1$	$\sum_{i=1}^{n} x_i^2$	[−100,100]	0
$f2$	$10^6 * x_i^2 + \sum_{i=2}^{n} x_i^2$	[−100,100]	0
$f3$	$\sum_{i=1}^{n} i * x_i^2$	[−100,100]	0
$f4$	$\frac{1}{4000} \sum_{i=1}^{n} x_i^2 - \prod_{i=1}^{n} \cos\left(\frac{x_i}{\sqrt{i}}\right) + 1$	[−600,600]	0
$f5$	$-20 \exp\left(-0.2 \sqrt{\frac{1}{n} \sum_{i=1}^{n} x_i^2}\right) + \exp\left(\frac{1}{n} \sum_{i=1}^{n} \cos(2\pi x_i)\right) + 20 + e$	[−32,32]	0
$f6$	$\frac{\pi}{n} \left\{ 10\sin^2(\pi y_i) + \sum_{i=1}^{n-1} (y_i - 1)^2 \left[1 + \sin^2(\pi y_{i+1})\right] + (y_n - 1)^2 \right\} + \sum_{i=1}^{n} u(x_i, 10, 100, 4)$	[−50,50]	0

Table 2 Result of the compared algorithms on ten dimensions

	*f*1			*f*2			*f*3		
	10	20	30	10	20	30	10	20	30
ABC	1.14E-16	4.27E-16	8.09E-16	1.21E-16	5.23E-16	9.61E-16	1.41E-16	4.73E-16	8.53E-16
	5.30E-17	1.11E-16	1.31E-16	5.36E-17	1.39E-16	2.95E-16	6.28E-17	1.08E-16	1.64E-16
GABC	7.00E-17	2.57E-16	5.37E-16	6.16E-17	2.48E-16	5.26E-16	7.54E-17	2.77E-16	5.23E-16
	1.40E-17	5.76E-17	8.88E-17	1.83E-17	5.19E-17	7.98E-17	2.96E-17	4.12E-17	7.81E-17
qABC	3.18E-17	5.41E-17	1.21E-16	2.95E-09	1.47E-18	1.56E-16	7.88E-17	2.51E-15	6.40E-16
	5.09E-17	4.82E-17	1.79E-16	1.61E-08	3.54E-18	8.51E-16	8.71E-17	1.22E-14	3.66E-16
OCABC	1.29E-55	6.41E-28	2.66E-20	8.88E-37	1.45E-43	1.50E-21	1.84E-37	4.49E-27	6.85E-19
	7.03E-55	3.43E-27	1.44E-19	4.85E-36	7.33E-43	8.11E-21	1.01E-36	2.46E-26	2.99E-18
ABC (ASF-MR)	7.00E-17	1.94E-16	2.63E-16	6.47E-17	1.91E-16	2.67E-16	6.26E-17	1.85E-16	2.64E-16
	1.35E-17	4.39E-17	2.71E-17	1.84E-17	4.71E-17	3.08E-17	2.02E-17	4.82E-17	3.19E-17
MABC	3.48E-57	1.15E-49	1.91E-47	6.70E-56	1.70E-49	9.21E-47	1.55E-56	1.42E-48	3.08E-46
	8.88E-57	2.32E-49	1.62E-47	1.74E-55	2.67E-49	1.62E-46	2.75E-56	3.68E-48	2.57E-46
Gaussian-ABC	1.70E-16	5.50E-16	1.08E-15	1.51E-16	1.17E-15	1.42E-14	1.43E-16	5.25E-16	1.15E-15
	7.63E-17	1.66E-16	3.03E-16	8.70E-17	3.28E-15	6.53E-14	6.11E-17	1.07E-16	3.89E-16
AABC	7.55E-17	2.65E-16	5.45E-16	7.29E-17	2.63E-16	5.20E-16	6.91E-17	2.53E-16	5.26E-16
	1.38E-17	3.66E-17	1.15E-16	1.60E-17	4.36E-17	1.01E-16	1.70E-17	3.53E-17	8.81E-17
STOC-ABC	4.08E-41	5.61E-40	7.33E-40	3.14E-38	1.60E-38	2.32E-38	1.49E-39	4.70E-39	1.22E-38
	1.18E-40	1.97E-39	9.05E-40	1.15E-37	7.89E-38	1.06E-37	6.81E-39	1.81E-38	2.54E-38
GLABC-pso	**2.62E-103**	**6.65E-93**	**8.69E-91**	**5.82E-101**	**2.20E-92**	**6.87E-90**	**1.30E-101**	**1.80E-92**	**1.91E-89**
	1.17E-102	**3.36E-92**	**1.71E-90**	**1.87E-100**	**8.71E-92**	**1.48E-89**	**4.51E-101**	**4.24E-92**	**3.73E-89**
GLABC-de	**2.50E-105**	**8.73E-99**	**2.73E-96**	**2.59E-102**	**3.87E-96**	**2.20E-92**	**8.66E-105**	**2.63E-97**	**1.56E-94**
	6.93E-105	**1.74E-98**	**3.88E-96**	**7.91E-102**	**2.06E-95**	**1.18E-91**	**1.64E-104**	**7.92E-97**	**4.52E-94**

(continued)

Table 2 (continued)

	*f*4			*f*5			*f*6		
D	10	20	30	10	20	30	10	20	30
ABC	8.59E-15	3.23E-14	6.72E-14	8.35E-17	3.92E-16	7.84E-16	1.16E-16	4.50E-16	8.77E-16
	3.28E-15	6.81E-15	1.25E-14	2.88E-17	9.13E-17	1.38E-16	4.92E-17	8.02E-17	1.48E-16
GABC	5.63E-15	2.29E-14	4.02E-14	4.97E-17	2.32E-16	4.65E-16	7.47E-17	2.68E-16	5.27E-16
	1.64E-15	3.63E-15	5.41E-15	1.81E-17	4.56E-17	6.63E-17	1.43E-17	3.76E-17	7.59E-17
qABC	**1.95E-15**	7.99E-15	**1.63E-14**	1.45E-18	9.98E-19	4.71E-18	4.51E-18	1.09E-17	2.44E-17
	2.70E-15	5.58E-15	**1.22E-14**	1.92E-18	1.20E-18	9.92E-18	6.53E-18	1.34E-17	3.23E-17
OCABC	2.55E-15	**2.66E-15**	3.88E-06	1.57E-32	7.15E-32	1.30E-06	1.35E-32	4.68E-32	2.71E-19
	6.49E-16	**0.00E+00**	2.13E-05	5.57E-48	3.05E-31	7.12E-06	5.57E-48	1.82E-31	1.37E-18
ABC (ASF-MR)	1.76E+01	1.89E+01	1.93E+01	3.97E+01	8.91E+05	1.18E+07	6.88E-17	8.77E+07	2.34E+08
	3.38E+00	3.58E+00	3.65E+00	5.45E+01	1.07E+06	1.00E+07	1.09E-17	3.73E+07	7.54E+07
MABC	5.03E-15	1.53E-14	2.95E-14	1.57E-32	1.57E-32	1.57E-32	1.35E-32	1.35E-32	1.35E-32
	1.70E-15	3.32E-15	3.05E-15	5.57E-48	5.57E-48	5.57E-48	5.57E-48	5.57E-48	5.57E-48
Gaussian-ABC	1.07E-14	3.96E-14	7.36E-14	1.68E-16	5.26E-16	2.42E-15	1.90E-16	5.26E-16	1.04E-15
	3.48E-15	9.26E-15	1.38E-14	2.12E-16	1.21E-16	7.68E-15	7.06E-17	1.46E-16	1.96E-16
AABC	3.61E-15	1.44E-14	3.42E-14	5.97E-17	2.60E-16	4.84E-16	7.64E-17	2.60E-16	4.74E-16
	1.60E-15	2.49E-15	5.50E-15	1.59E-17	3.70E-17	1.14E-16	2.64E-17	4.54E-17	1.02E-16
STOC-ABC	1.05E-14	4.25E-14	9.27E-14	1.57E-32	1.64E-32	3.95E-31	1.35E-32	1.88E-32	5.01E-32
	3.66E-15	8.31E-15	2.91E-14	5.57E-48	2.24E-33	2.04E-30	5.57E-48	1.11E-32	4.73E-32
GLABC-pso	*5.63E-15*	*1.82E-14*	*3.37E-14*	**1.57E-32**	**1.57E-32**	3.46E-03	**1.35E-32**	**1.35E-32**	**1.35E-32**
	1.35E-15	*4.52E-15*	*4.47E-15*	**5.57E-48**	**5.57E-48**	1.89E-02	**5.57E-48**	**5.57E-48**	**5.57E-48**
GLABC-de	*5.15E-15*	*1.78E-14*	*3.06E-14*	**1.57E-32**	**1.57E-32**	**1.57E-32**	**1.35E-32**	**1.35E-32**	**1.35E-32**
	1.66E-15	*3.95E-15*	*3.46E-15*	**5.57E-48**	**5.57E-48**	**5.57E-48**	**5.57E-48**	**5.57E-48**	**5.57E-48**

The first column illustrates the average, and the second column is the standard deviation in parentheses.

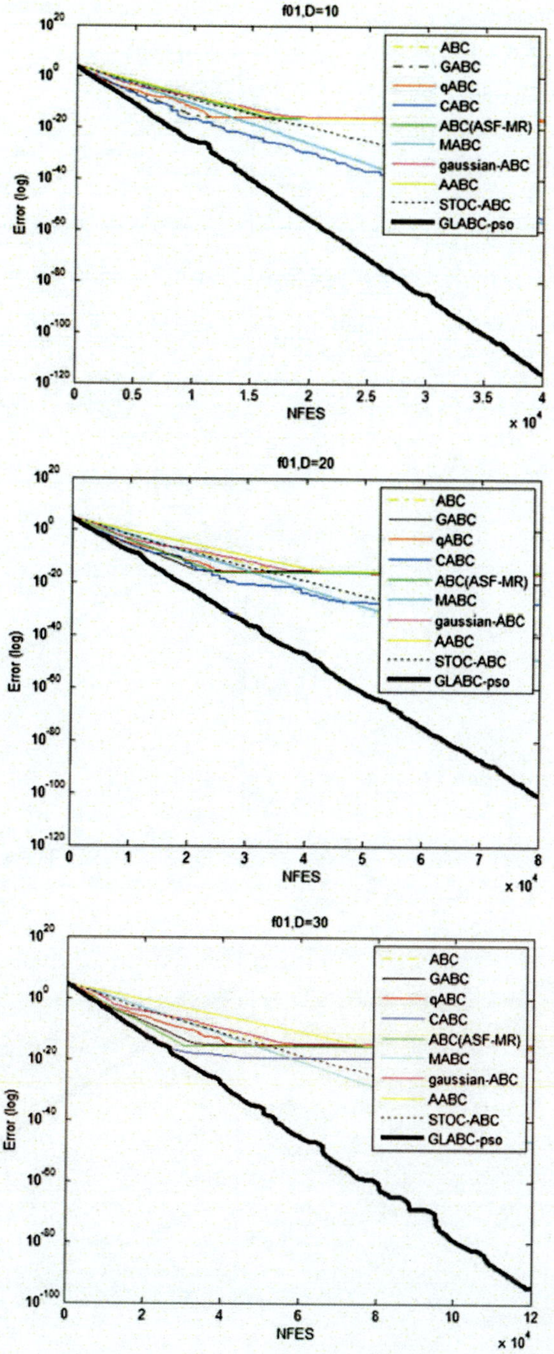

Fig. 2 The convergence comparison on Fun.1 and 3 with 10, 20, and 30 dimensions

Fig. 2 (continued)

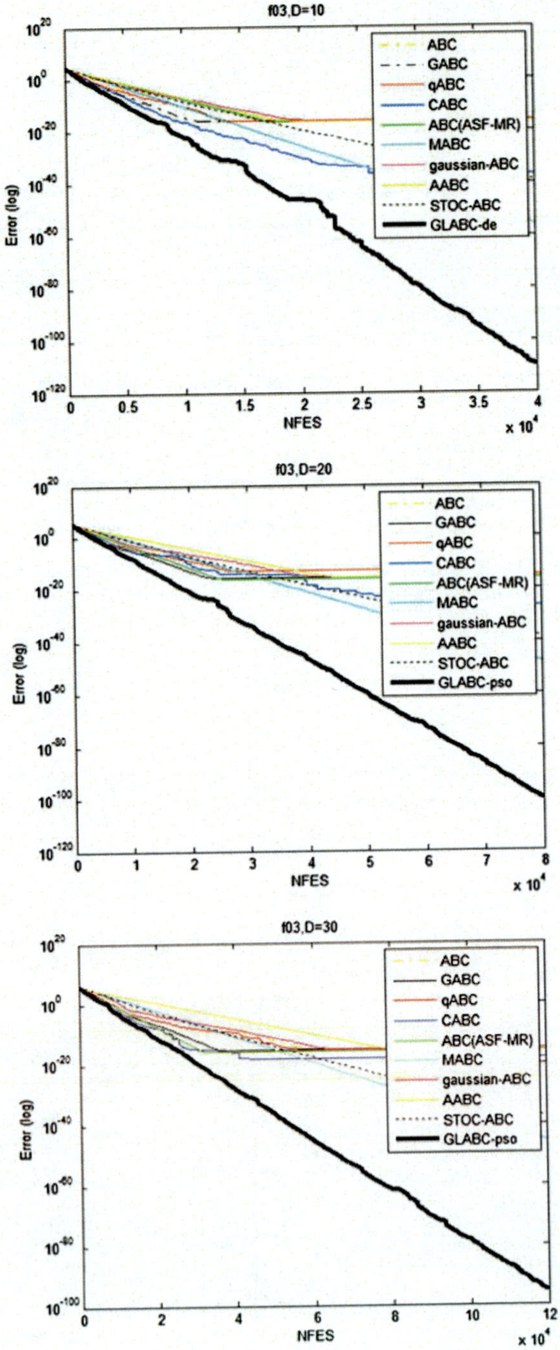

to the potential global optimum, and the global-best leading strategy in onlooker bees also enables the individual bees to make a local search, thus resulting in getting more precise solution accuracy.

For the multimodal problems, the proposed two GLABC algorithms achieved superior or similar performance on $f9$–$f12$, where the number of local optima increases exponentially with the problem dimension. It should be mainly owing to the Gaussian distribution operators F and M enable bees have relative power search ability to jump out of the local optima, which should maintain the global search abilities for the onlooker and employed bees (Fig. 2).

4 Conclusion

This chapter first analyzes the performance of the search strategy in the standard ABC by comparing with the PSO and the DE. Then, by introducing the advantage of PSO and DE into the employed bee phase and utilizing a global-best leading strategy for onlooker bee phase, this chapter proposes two new algorithms, GLABC-pso and GLABC-de, to accelerate the convergence speed and improve the solution accuracy on the conditions of maintaining the global search ability. Finally, the two proposed algorithms are applied to solve the large-scale transmission pricing problem. Numerical experiment results have demonstrated the effectiveness and efficiency of the two GLABC algorithm. Future study will focus on further improving the performance of the standard ABC algorithm and applying it to real-word problems.

Acknowledgments The authors acknowledge the support from the National Natural Science Foundation of China (No. 61571236), the Science and Technology on Space Intelligent Control Laboratory (KGJZDSYS-2018-02), the Research Committee of University of Macau (MYRG2015-00011-FST, MYRG2018-00035-FST), the Science and Technology Development Fund of Macau SAR under Grant 041-2017-A1, and Postgraduate Research and Practice Innovation Program of Jiangsu Province (SJCX18_0300, KYCX18_0929).

References

1. D. Karaboga, An idea based on Honey Bee Swarm For Numerical Optimization, Erciyes Univ., Kayseri, Turkey, Tech. Rep.-TR06, (2005)
2. S.K. Goudos, K. Siakavara, J.N. Sahalos, Novel spiral antenna design using artificial bee colony optimization for UHF RFID applications. IEEE Antennas Wirel. Propag. Lett. **13**, 528–531 (2014)
3. X. Li, M. Yin, Hybrid differential evolution with artificial bee colony and its application for design of a reconfigurable antenna array with discrete phase shifters. IET Microwaves Antennas Propag. **6**(14), 1573–1582 (2012)
4. X. Zhang, X. Zhang, S.L. Ho, et al., A modification of artificial bee colony algorithm applied to loudspeaker design problem. IEEE Trans. Magn. **50**(2), 737–740 (2014)
5. X. Zhang, X. Zhang, S.Y. Yuen, et al., An improved artificial bee colony algorithm for optimal design of electromagnetic devices. IEEE Trans. Magn. **49**(8), 4811–4816 (2013)

6. C. Ozturk, D. Karaboga, Hybrid artificial bee colony algorithm for neural network training[C]//Evolutionary Computation (CEC), in *2011 IEEE congress on*. IEEE, (2011), pp. 84–88
7. T.J. Hsieh, H.F. Hsiao, W.C. Yeh, Forecasting stock markets using wavelet transforms and recurrent neural networks: An integrated system based on artificial bee colony algorithm. Appl. Soft Comput. **11**(2), 2510–2525 (2011)
8. H. Xu, M. Jiang, K. Xu, Archimedean copula estimation of distribution algorithm based on artificial bee colony algorithm. J. Syst. Eng. Electron. **26**(2), 388–396 (2015)
9. P.W. Tsai, M.K. Khan, J.S. Pan, et al., Interactive artificial bee colony supported passive continuous authentication system. IEEE Syst. J. **8**(2), 395–405 (2014)
10. Q.K. Pan, L. Wang, K. Mao, et al., An effective artificial bee colony algorithm for a real-world hybrid flowshop problem in steelmaking process. IEEE Trans. Autom. Sci. Eng. **10**(2), 307–322 (2013)
11. S.C. Horng, Combining artificial bee colony with ordinal optimization for stochastic economic lot scheduling problem. IEEE Trans. Syst. Man Cybern. Syst. **45**(3), 373–384 (2015)
12. H. Duan, S. Li, Artificial bee colony??? Based direct collocation for reentry trajectory optimization of hypersonic vehicle. IEEE Trans. Aerosp. Electron. Syst. **51**(1), 615–626 (2015)
13. M. Li, H. Zhao, X. Weng, et al., Artificial bee colony algorithm with comprehensive search mechanism for numerical optimization. J. Syst. Eng. Electron. **26**(3), 603–617 (2015)
14. W. Gao, S. Liu, A modified artificial bee colony algorithm. Comput. Oper. Res. **39**(3), 687–697 (2012)
15. G. Zhu, S. Kwong, Gbest-guided artificial bee colony algorithm for numerical function optimization. Appl. Math Comput. **217**(7), 3166–3173 (2010)
16. D. Karaboga, B. Gorkemli, A quick artificial bee colony (qABC) algorithm and its performance on optimization problems[J]. Appl. Soft Comput. **23**, 227–238 (2014)
17. W. Gao, S. Liu, L. Huang, A global best artificial bee colony algorithm for global optimization. J. Comput. Appl. Math. **236**(11), 2741–2753 (2012)
18. W. Gao, S. Liu, L. Huang, A novel artificial bee colony algorithm based on modified search equation and orthogonal learning. IEEE Trans. Cyber. **43**(3), 1011–1024 (2013)
19. B. Akay, D. Karaboga, A modified artificial bee colony algorithm for real-parameter optimization[J]. Inform. Sci. **192**, 120–142 (2012)
20. Y. Shi, R. Eberhart, A modified particle swarm optimizer, in *IEEE World Congress on Computational Intelligence*, (1998), pp. 69–73
21. R. Storn, K. Price, Differential evolution-a simple and efficient huristic for global optimization over continuous spaces. J. Global Optmi. **11**(4), 341–359 (1997)
22. R.A. Krohling, Gaussian particle swarm with jumps. IEEE Congr. Evol. Comput. **2**, 1226–1231 (2005)
23. X. Yao, Y. Liu, G. Lin, Evolutionary programming made faster. IEEE Trans. Evol. Comput. **3**(2), 82–102 (1999)
24. M.M. Ali, C. Khompatraporn, Z.B. Zabinsky, A numerical evaluation of several stochastic algorithms on selected continuous global optimization test problems. J. Glob. Optim. **31**(4), 635–672 (2005)
25. X. Liao, J. Zhou, R. Zhang, et al., An adaptive artificial bee colony algorithm for long-term economic dispatch in cascaded hydropower systems. Int. J. Electr. Power Energy Syst. **43**(1), 1340–1345 (2012)
26. R.C. Blair, J.J. Higgins, A comparison of the power of wilcoxon's rank-sum statistic to that of student'st statistic under various nonnormal distributions. J. Educ. Behav. Stat. **5**(4), 309–335 (1980)
27. L.D.S. Coelho, P. Alotto, Gaussian artificial bee colony algorithm approach applied to Loney's solenoid benchmark problem. IEEE Trans. Magn. **47**(5), 1326–1329 (2011)
28. R. Lu, H.D. Hu, M.L. Xi, et al., An improved artificial bee colony algorithm with fast strategy and its application. Comput. Electr. Eng. **78**, 79–88 (2019)

Intelligent Control of Ultrasonic Motor Using PID Control Combined with Artificial Bee Colony Type Neural Networks

Shenglin Mu, Satoru Shibata, Tomonori Yamamoto, Shota Nakashima, and Kanya Tanaka

1 Introduction

Ultrasonic motor, which can be called short for USM, is a kind of electronic motor driven by frictions generated according to the ultrasonic vibrations on piezoelectric materials [1, 2]. Comparing with traditional electromagnetic motors, USMs are considered as a novel kind of actuators because they are commercialized since 1980s. Owing to the excellent features on USM, such as compactness in size, low mass, quiet operation, high retention torque, and no electromagnetic noise, etc., they are widely applied in various applications. USMs are used as actuators in watches, auto-focusing cameras, industrial and robotic manipulators, and medical devices [3, 4]. However, owing to the USM's friction driven principle there are disadvantages such as short life duration, complexity in driver design, no mathematic model for control, characteristic changes, and nonlinearity [3, 5].

Therefore, in recent years, owing to the development of intelligent algorithms [6–9] various kinds of researches have been proposed for the intelligent control

S. Mu (✉) · S. Shibata
Graduate School of Science and Engineering, Ehime University, Matsuyama, Japan
e-mail: mu.shenglin.du@ehime-u.ac.jp; shibata.satoru.mg@ehime-u.ac.jp

T. Yamamoto
Faculty of Collaborative Regional Innovation, Ehime University, Matsuyama, Japan
e-mail: yamamoto.tomonori.mh@ehime-u.ac.jp

S. Nakashima
Graduate School of Science and Engineering, Yamaguchi University, Yamaguchi, Japan
e-mail: s-naka@yamaguchi-u.ac.jp

K. Tanaka
School of Science and Technology, Meiji University, Kanagawa, Japan
e-mail: kanya728@meiji.ac.jp

Y. Li, H. Lu (eds.), *3rd EAI International Conference on Robotic Sensor Networks*, EAI/Springer Innovations in Communication and Computing,
https://doi.org/10.1007/978-3-030-46032-7_7

of USMs compensating the characteristics and nonlinearity. In this section, several schemes using intelligent algorithms are introduced. In [3], a neural network-based adaptive control strategy is proposed for compensating the nonlinearity of servo motor. In the proposed method, two neural networks have been adopted for system identification (NNI) and control (NNC), respectively. In [10], a new position control scheme for ultrasonic motor using NN was proposed. In the proposed scheme, NN based on general back-propagation (BP) algorithm was applied. A neural network scheme combined with Particle Swarm Optimization (PSO) algorithms was proposed in previous research [11]. The proposed method uses PSO algorithms for easier estimation in NN's learning without considering Jacobian. The application of swarm intelligence algorithm was confirmed effective. In this research, to obtain satisfactory control performance in position control of USM, we introduce a novel swarm intelligence algorithm of ABC. The algorithm is developed by D. Karaboga, inspired by the behavior of honey bees [12, 13]. It is useful with simplicity and effectiveness in optimizations. In this research, we introduce it into the learning of NN on PID control of USM. The effectiveness of the proposed scheme is confirmed by position control experiments on an existing USM servo system.

The paper is organized as follows. In Sect. 2, the PID control for USM is introduced. The proposed method using ABC algorithm is introduced in Sect. 3. The experiments on the servo system are introduced in Sect. 4. Finally, the paper is concluded in Sect. 5.

2 PID Control for USM

PID (Proportional-Integral-Derivative) control is a popular control algorithm which is very widely used in various of industrial fields. Its simplicity and effectiveness have been verified in innumerable applications. It is introduced to USM control because that it works quite well even without a mathematical model of plant. The PID control scheme for USM can be designed as shown in Fig. 1. In the block diagram, $G_{PID}(z^{-1})$ represents the incremental PID controller for the USM. In the scheme, $r(k)$, $u(k)$, and $y(k)$ are the objective input, the control input, and the output in discrete time, respectively. $e(k)$ is the error between the objective input and the output as shown in the following equation.

$$e(k) = r(k) - y(k). \tag{1}$$

Fig. 1 Block diagram of PID control for USM

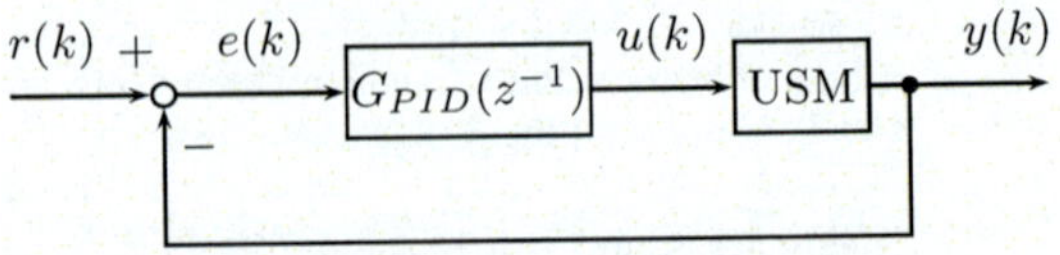

The control input of the system can be synthesized as

$$u(k) = u(k-1)+(K_P+K_I+K_D)e(k)-(K_P+2K_D)e(k-1)+K_De(k-2). \quad (2)$$

The PID controller can be denoted as follows.

$$G_{PID}(z^{-1}) = \frac{K_P(1-z^{-1}) + K_I + K_D(1-z^{-1})^2}{1-z^{-1}}. \quad (3)$$

In the equation, K_P is the proportional gain, K_I is the integral gain, K_D is the derivative gain, respectively. By tuning to obtain proper gains in the PID control, satisfactory control performance can be achieved even without a mathematical model of the USM. To obtain the gains of PID control, there are famous tuning methods for the PID controller, such as Ziegler–Nichols rule and Cohen–Coon rule. But, for the USM control with the characteristics change and nonlinearity, the tuning of the three parameters is quite time consuming and difficult. In this research, an intelligent control method is introduced to obtain the gains automatically.

3 Proposed Method

3.1 Proposed Intelligent PID Control Using ABC

The proposed intelligent method is as Fig. 2 shows. The scheme is designed for tuning PID gains automatically by the estimation of NN to achieve good performance in USM control. In the NN controller, the topologic structure is designed as Fig. 3 shows. There are $i = 3$ neurons in the input layer with the input signals of discrete time errors which can be expressed as

$$I_i(k) = \{e(k), e(k-1), e(k-2)\}. \quad (4)$$

There are $j = 6$ neurons in the hidden layer. $w_{ij}(k)$ means the weights between the input layer and the hidden layer in discrete time. The sum estimations of net_{ij} in them are implemented by the following equations.

$$net_{ij} = \sum_{i=1}^{3} w_{ij}(k)I_i(k). \quad (5)$$

The output of hidden neurons $H_j(k)$ will be estimated by the activation function. In this research, the hyperbolic tangent function shown in is applied as the activation function.

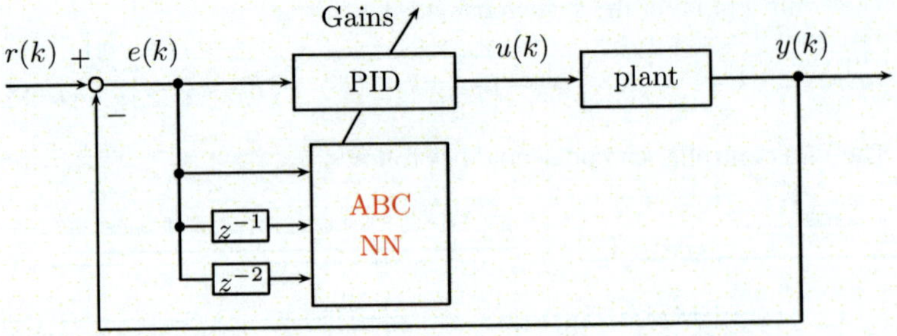

Fig. 2 Block diagram of the proposed intelligent PID control for USM

Fig. 3 Topologic structure of NN in the proposed scheme

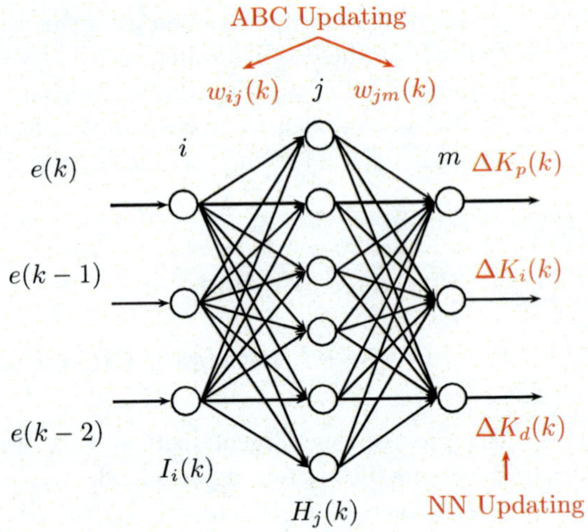

$$H_j(k) = \frac{1 - e^{(-net_{ij})}}{1 + e^{(-net_{ij})}}. \tag{6}$$

In the neurons of output layer, there are $m = 3$ neurons. The sum estimations are implemented as

$$net_{jm} = \sum_{j=1}^{6} w_{jm}(k) H_j(k) \tag{7}$$

where $w_{jm}(k)$ means the weights between the input layer and the hidden layer in discrete time. The output of the neurons $O_m(k)$ can be estimated as

$$O_m(k) = \frac{1 - e^{(-net_{ij})}}{1 + e^{(-net_{jm})}} \tag{8}$$

The output signals of NN are applied as the variation of PID gains

$$
\begin{aligned}
\Delta K_P(k) &= O_1(k) \\
\Delta K_I(k) &= O_2(k) \\
\Delta K_D(k) &= O_3(k).
\end{aligned}
\tag{9}
$$

Then the gains of PID control can be estimated as

$$
\begin{aligned}
\bar{K}_P(k) &= K_P(k-1) + \Delta K_P(k) \\
\bar{K}_I(k) &= K_I(k-1) + \Delta K_I(k) \\
\bar{K}_D(k) &= K_D(k-1) + \Delta K_D(k).
\end{aligned}
\tag{10}
$$

There for the output of PID controller in discrete time can be updated according to the variation of gains.

$$
\begin{aligned}
u(k) = u(k-1) &+ (\bar{K}_P(k) + \bar{K}_I(k) + \bar{K}_D(k))e(k) \\
&-(\bar{K}_P(k) + 2\bar{K}_D(k))e(k-1) + \bar{K}_D(k)e(k-2).
\end{aligned}
\tag{11}
$$

3.2 *Artificial Bee Colony*

ABC is a global optimization algorithm proposed by Karaboga in 2005. It is inspired from the food foraging behavior of honey bees. It is an important algorithm of swarm intelligence which performs well in optimization problems with no derivative information required beforehand [12].

In the standard model of ABC algorithm, the food source for bees are considered as candidate solutions in optimization problem. Generally, three types of bees are modeled in ABC algorithm, the employed, the onlooker, and the scout bees. ABC colony consists equal number of employed and onlooker bees. Employed bees explore search space for food according to the information in their memories. Onlooker bees obtain the information from employed bees in the hive to select food sources for further extraction of nectar. If the nectar quantity in food source is low or exhausted, then scout bee randomly finds a new food source in search space [13].

In our proposed method, ABC is introduced to NN to update the weights in the learning process. To realize the ABC algorithm in the weight updating process, weight solutions in the proposed scheme are pre-processed as a one-dimensional vector as follows.

$$
\mathbf{W}_n = \{w(k)_{n,1}, w(k)_{n,2}, \cdots, w(k)_{n,D}\}
\tag{12}
$$

where $n = (1, 2, \cdots, N)$ is the index of food sources. $D = i \times j + j \times m$ is the dimension of solutions. In the topologic scheme of the proposed NN scheme shown in Fig. 3, there are 36 weights to be updated. A one- dimensional solution containing 36 weight values is considered to be optimized by the ABC scheme we constructed. Then, the weights are initialized as

$$\mathbf{W}_n = \mathbf{W}_{min,n} + rand(0, 1)(\mathbf{W}_{max,n} - \mathbf{W}_{min,n}) \tag{13}$$

where $\mathbf{W}_{min,n}$ and $\mathbf{W}_{max,n}$ are the minimum and maximum values for the weights. $rand(0, 1)$ is the random number between 0 and 1.

According to the process in ABC the possible solutions are updated by different type of bees. Firstly, the employed bees search the possible space around the solution offering a modification on current solution.

$$\mathbf{V}_n = w(k)_{p,n} + \phi_{p,n}(w(k)_{p,n} - w_{q,n}) \tag{14}$$

where $\phi_{p,n}$ is a random number in the range of $[-1, 1]$. If a new solution is with higher fitness in the searching process, the current solution will be replaced by the new on. If the solution's fitness is not higher than the current one, the current solution will be kept. Onlooker bees: chooses a solution $\mathbf{W}_g$ to be updated by $\mathbf{V}_g$ depending on the probability $prob_g$

$$prob_g = \frac{fitness_g}{\Sigma_{h=1}^{N} fitness_h}. \tag{15}$$

Scout bees generate a novel when the solution cannot be improved in predetermined cycles. In the proposed scheme, the weights will be updated according to the evaluation on $e(k)$. Therefore, the fitness for evaluation will be estimated as

$$fitness(k) = \frac{1}{1 + e(k)^2}. \tag{16}$$

When the error signal is 0, the weights get highest fitness of 1 for evaluation. When the error signal gets larger, the fitness will deteriorate approaching to 0. In this research, weights will be updated by the proposed ABC algorithm, converging to minimize the error $e(k)$ automatically.

4 Experiments

To confirm the effectiveness of the proposed method in this research, a group of position control experiments have been implemented on an existing USM servo system. Figure 4 shows the photograph of the USM servo system. The USM is set on the left side offering the horizontal rotation. A magnetic brake is set on the

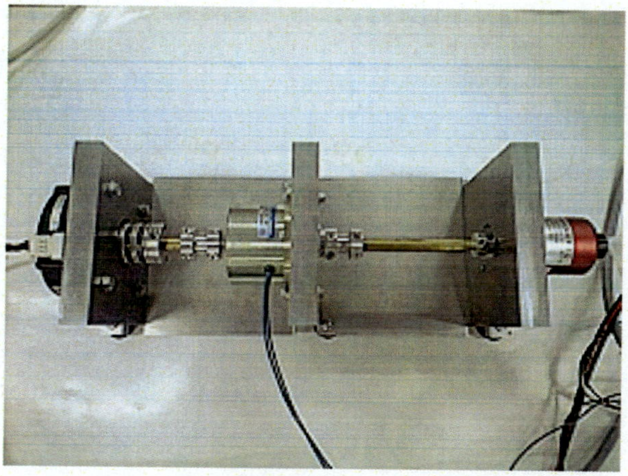

Fig. 4 Photograph of USM servo system

Table 1 Specifications of USM

Driving frequency	50 kHz
Rated torque	0.5 N·m
Rated speed	250 rpm
Holding (maximum) torque	0.1 N·m
Response time	Less than 1 ms
Endurance time	2000 h

middle of the axis, and the encoder with pink color is connected on the other end on the axis. The position information fetched from the encoder is input to the counter board set up in the PC. Information of the control input calculated from the output and the target value in the PC is transferred to the driving circuit via I/O board. Table 1 shows the performance of the USM, the encoder, and the load.

Frequency control is implemented in the servo system. The input-output characteristic between control frequency and velocity is shown in Fig. 5. It is shown that the USM gets maximum rotating velocity when the driving frequency is about 40.5 kHz, and it gets stop when the frequency is around 46.8 kHz. To obtain satisfactory control performance, the relatively linear range of control frequency [42, 46.8] kHz is used as driving frequency.

In order to validate the effectiveness of the proposed method, the position control experiments with an objective input of step signal with the amplitude of 45 degree was applied. The period of the input is 2 second. Number of employ bees and onlooker bees are both set as 20. The cycle limit which decides when the scout bees appear is set as 20. When the evaluation is not improved in 20 cycles the solution will be replaced by the solutions generated by scout bees. In the initialization, the maximum and minimum values of weights are +1 and −1. To confirm the effectiveness, a fixed gain type PID with the gains of $K_P = 5.0$, $K_I = 1.0$, $K_D = 0.1$ is introduced for comparison.

The step response of the proposed scheme is shown in Fig. 6. The right figure gives the response in position control. It is clear that the positioning in rotation

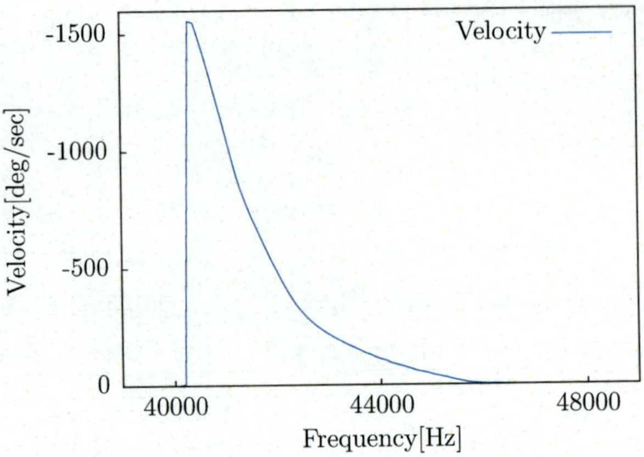

Fig. 5 Characteristics of USM servo system

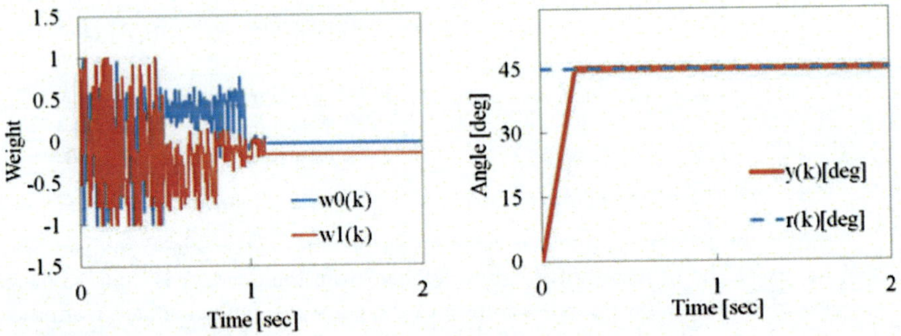

Fig. 6 Step response of PID control for USM

of USM is quick and precise with no obvious overshoot or delay. There is almost no steady-state error in the position control. The variation of two representative weights is shown in the left figure in Fig. 6. It can be seen that the weights varied at the first second in the rising up. The oscillation happened at the initial phase within the first second, then the value converged to certain range. To see the control performance clearly, the enlarged view of steady-state is shown in Fig. 7. The figure on the left side shows the response of the proposed intelligent scheme. It is clear that when the USM rotates to the position around the objective one at +45 degree, there was oscillation around the objective position. The amplitude of the oscillation was within the range of 0.25 degree. After the oscillation, the rotation converged to the objective position without obvious error. The figure on the right side shows the step response using the fixed gain type PID control with the gains of $K_P = 5.0$, $K_I = 1.0$, $K_D = 0.1$. We can see obvious overshoot when the rotation approach to the objective input. The oscillation of the fixed gain type PID control lasted until the end of the experiment. It is clear that the proposed intelligent scheme is with higher

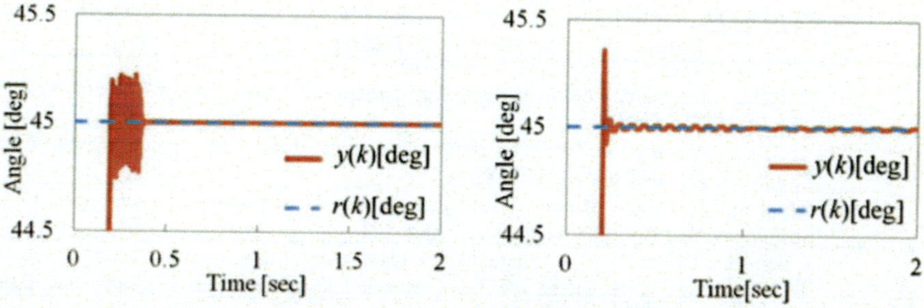

Fig. 7 Enlarged view of step response of PID control for USM

Fig. 8 Variation of fitness in the proposed scheme for USM

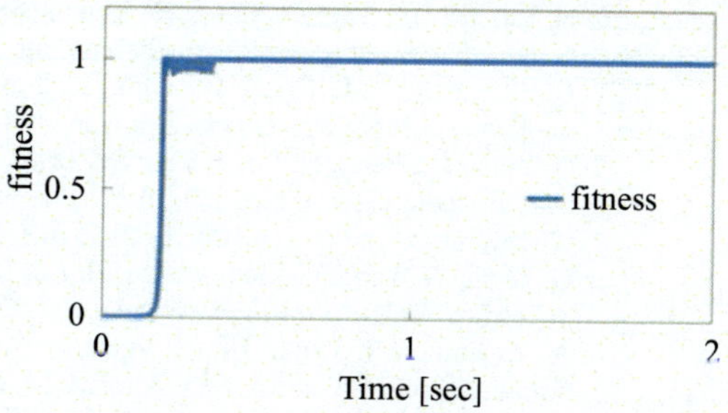

control performance in position control than the fixed type PID control. The fitness in the proposed scheme is shown in Fig. 8. It can be seen clearly that the evaluation of the solutions approached to higher values in a short time. The weight updated of the proposed ABC type NN converged to the almost maximum fitness 1 within 0.5 second in the response.

5 Conclusions

In this research, a novel swarm intelligence algorithm of ABC is introduced to construct an intelligent PID control method for obtaining satisfactory control performance of USM. The algorithm is introduced in the learning of NN type PID control for USM. The effectiveness of the proposed scheme is confirmed by position control experiments on an existing USM servo system. The proposed method gives good steady-state response in the position control. The proposed method can be used without the consideration of the Jacobian estimation in traditional BP type neural network in the intelligent PID control. According to the experimental study, it is confirmed that the proposed scheme is useful with simplicity and effectiveness.

References

1. T. Kenjo, T. Sashida, *An Introduction of Ultrasonic Motor* (Oxford Science Publications, Oxford, 1993)
2. K. Adachi, *Actuator with Friction Drive: Ultrasonic Motor*. The Japan Society of Mechanical Engineers, vol. 108, pp. 48–51, 2005
3. C. Zhao, *Ultrasonic Motor - Technologies and Applications* (Science Press Beijing and Springer-Verlag Berlin Heidelberg, Beijing, 2011)
4. I. Yamano, T. Maeno, Five-fingered robot hand using ultrasonic motors and elastic elements, in *Proceedings of the 2005 IEEE International Conference on Robotics and Automation*, pp. 2673–2678, 2005
5. T. Senjyo, H. Miyazato, K. Uezato, *Quick and Precise Position Control of Ultrasonic Motor using Hybrid Control, Electrical Engineering in Japan*, vol. 116, No. 3, pp. 83–95, 1996
6. H. Lu, Y. Li, S. Mu, D. Wang, H. Kim, S. Serikawa, Motor anomaly detection for unmanned aerial vehicles using reinforcement learning. IEEE Internet Things J. **5**(4), 2315–2322 (2018)
7. H. Lu, Y. Li, M. Chen, H. Kim, S. Serikawa, Brain Intelligence: go beyond artificial intelligence. Mobile Networks Appl. **23**, 368–375 (2018)
8. Q. Zhou, J. Cheng, H. Lu, Y. Fan, S. Zhang, X. Wu, B. Zheng, W. Ou, L.J. Latecki, Learning adaptive contrast combinations for visual saliency detection. Multimed. Tools Appl., 1–29 (2018). https://doi.org/10.1007/s11042-018-6770-2
9. Q. Zhou, W. Yang, G. Gao, W. Ou, H. Lu, J. Chen, L.J. Latecki, *World Wide Web* vol. 22(2), pp. 555–570 (2019). https://doi.org/10.1007/s11280-018-0556-3
10. A. Aoyama, F.J. Doyle III, V. Venkatasubramanian, A fuzzy neural-network approach for nonlinear process control. Eng. Appl. Artif. Intell. **8**, 483–498 (1995)
11. S. Mu, K. Tanaka, S. Nakashima, D. Alrijadjis, Real-time PID Controller using Neural Network Combined with PSO for ultrasonic motor. ICIC Express Lett. **8**(11), (2014)
12. K. Karaboga, An idea based on Honey Bee Swarm for numerical optimization. Technical report-tr06, Erciyes university, engineering faculty, 2005
13. D. Karaboga, B. Gorkemli, C. Ozturk, N. Karaboga, A Comprehensive survey: Artificial Bee Colony (ABC) algorithm and applications. Artif. Intell. Rev. **42**, 21–57 (2014)

An Adaptable Feature Synthesis for Camouflage

Guangxu Li, Ziyang Yu, Jiying Chang, and Hyoungseop Kim

1 Introduction

Recently a trend towards digital camouflage has appeared in military forces, which is not due to instinct creativity but scientific fact. The pixels printed on the military supplies are better at mimicking fractal patterns, which interprets the human visual system to fixate on ın the pattern of white noise. Generally, the digital camouflage has two layers: a micropattern, which describes the details of surface in pixel level; and a macropattern, by which the shapes are better formed. The reasonable combination of these two layers makes the invisibility of targets insensitive from the distance [1].

The traditional manual design of camouflage texture mainly depends on the designer's experience, which were inspired by natural environments and based on biological or psychological principles such as blending and disruption. According to the spatial color mixing principle, digital camouflage texture is formed by pixels

G. Li (✉)
School of Electronics and Information Engineering, Tiangong University, Tianjin, China

Tianjin Key Laboratory of Optoelectronic Detection Technology and System, Tianjin, China
e-mail: liguangxu@tjpu.edu.cn

Z. Yu
School of Electronics and Information Engineering, Tiangong University, Tianjin, China

J. Chang
School of Precision Instruments and Optoelectronics Engineering, Tianjin University, Tianjin, China

H. Kim
Department of Control Engineering, Kyushu Institute of Technology, Fukuoka, Japan
e-mail: kim@cntl.kyutech.ac.jp

Y. Li, H. Lu (eds.), *3rd EAI International Conference on Robotic Sensor Networks*, EAI/Springer Innovations in Communication and Computing,
https://doi.org/10.1007/978-3-030-46032-7_8

"lattice," which is easier to be blended with the natural background. In [2] Kilian and Hepfinger first time proposed computer aided camouflage texture devaluation. The scope of evaluation is limited to target acquisition reduction performance of camouflaged uniforms against the reconnaissance. A set of features for each component in the CIELAB color space is used for the texture measures and first-order descriptive statistics over specified regions in a scene as functions of orientation and scale.

Generally, the computer aided camouflage design is illuminated in Fig. 1. The battlefield images are collected to generate the templates of camouflage at the first. The pixel lattices are sampled on the template images and reform into the texture patterns. Finally, map the pattern to the military objects and adjust it according to the on-set tests. According to a serious of experimental tests, Hogervorst et al. [3] validated that the camouflage ability of a specialized pattern is better or comparative to a universal pattern. The multiple disciplines of tomograph processing, computer vision, dynamics, ergonomics as well as human visual perception have been integrated into the design considerations of camouflage. In [4], the researchers establish a "dazzle camouflage" model to solve the most attempts disruption at concealment caused by motion. Lin et al. [5] proposed an assessment method based on the differences in terms of detectability and discriminability of the camouflage patterns. They improved that the camouflage patterns could significantly affect sensitivity of recognition. Xue et al. improved a saliency map of pattern template to reflect the hide quality and camouflage performance in a certain extent [6]. In the later works, they add the clustering for extraction of the major color of background camouflage texture. And a greedy algorithm is applied to optimize the distribution of texture template to form the macropattern of texture [7]. How to provide an automatically camouflage design which adapts for the various battlefields quickly is the most challenging aspect. In this paper, we propose to use feature transitions based approach to extract the major characters of the battlefield background images. In order to simulate the efficiency of observation in the different distances, we separate the background images by scales presentation. The global impress of camouflage is extracted from distant view images, well its texture is depicted using the close short images. Moreover, we extent the feature transitions to the 3D surfaces applying for the extensive military installations.

2 Feature Transitions

2.1 Related Works

Feature transitions under the deep neural networks framework noted as "style transitions" have been widely used since the Gatys et al. novel works [9]. They use the neural networks to capture the style of artistic images and transfer them to real-world photographs. The high-level feature representations of images are obtained

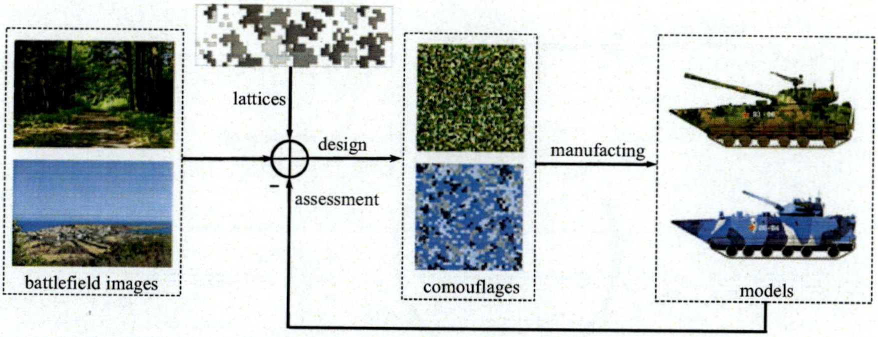

Fig. 1 Outline of camouflage design procedure

from the VGG convolutional network to separate and recompose content and style. A new neurally activated image is generated by the optimization model, which looks similar to texture images, and feature correlations as the feature image. After that, the follow-up researches consider different ways to represent the feature in neural networks. In [8] the authors tried features conversion algorithm of Gatys' work. Several variations of the manner in which image features applicable to different networks represent the archiving of different objects (such as lighting or season shifts) are proposed. The works of [10] extend the artistic feature works to the video making. They propose a new initialization and loss functions applicable which could transfer the feature from one image to a video sequence. Lu et al. [13–15] propose a framework to extract the features of all-weather environment using convolutional network. They certified that feather consistency of surrounding could be kept after network training. However, for the battlefield application, the camouflage generated from few images is easily to be one-sided. A CNN network with ability of dealing with a set of images shorted in a large scale is necessary for this work.

2.2 *Feature Transfer for Camouflage*

The purpose of feature conversion is to generate a feature image s, denoted as consisting of a sequential original image im. It could be developed as a problem of energy minimization consisting of a set of losses. The key idea is to extract the features of all input images through a convolutional neural network and then encode the feature based on the relevance of these features. The framework of this network demonstrated in Fig. 2. The feature loss is also a mean squared error. And the filter responses are expressed using the Gram matrices. There is $S^l \in \mathbb{R}^{N_l \times N_l}$ for the feature images s and $G^l \in \mathbb{R}^{N_l \times N_l}$ for the image im. These are calculated as $\tilde{S}^l_{ij} = \sum_{k=1}^{M_l} S^l_{ik} S^l_{jk}$ and $\tilde{F}^l_{ij} = \sum_{k=1}^{M_l} F^l_{ik} F^l_{jk}$. Let $\mathcal{L}_{fea}$ be the set of layers that uses to represent the feature, then we have the feature loss:

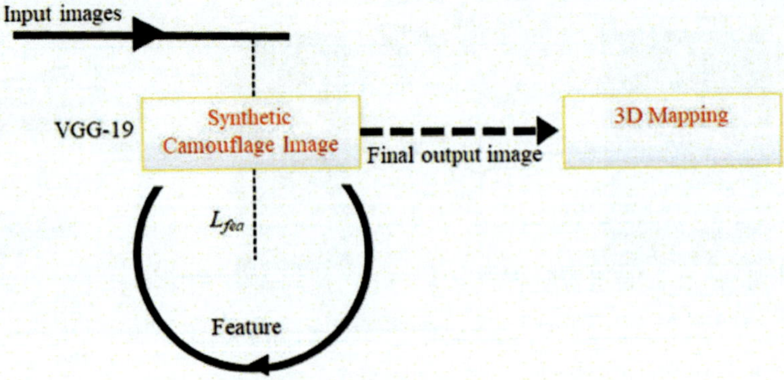

Fig. 2 Framework of CNN for camouflage generation

$$\mathcal{L}_{fea}(s, im) = \sum_{l \in \mathcal{L}_{fea}} \frac{1}{N_l^2 M_l^2} \sum_{i,j} \left(G_{i,j}^l - S_{i,j}^l\right)^2. \tag{1}$$

After training the feature image, we print a texture image onto the final output image. If Φ^l is defined to represent the function implemented by the convolutional network from the input layer to the layer l. Then the feature maps could be extracted by the network from the original image c, the feature image s and the featured image im. If denote $C^l = \Phi^l(c)$, $S^l = \Phi^l(s)$ and $M^l = \Phi(im)$, respectively, the dimensions of these feature maps are represented by $N_l \times M_l$, where N_l is the number of filters in the layer and M_l is the spatial dimension of the feature map, which is the product of width and height. We denoted the texture loss L_{tex} as the mean squared error between $C^l \in \mathbb{R}^{N_l \times M_l}$ and $F^l \in \mathbb{R}^{N_l \times M_l}$. Let L_c be the set of layers to be used for texture representation, then we have:

$$\mathcal{L}_{tex}(c, im) = \sum_{l \in \mathcal{L}_c} \frac{1}{N_l M_l} \sum_{i,j} \left(M_{i,j}^l - C_{i,j}^l\right)^2. \tag{2}$$

3 Feature Transfer to 3D Surface

3.1 Overview

The technique maps an image to a 3D surface is well-known the texture mapping. Parameterization of polygonal geometric surfaces has direct use for this purpose, which associates a coordinate in image space with every location on the surface of a geometric object. In order to normalize the transformation and keep the surface image adaptively transformed, the surface curvature map was taken into account. In addition, we introduce the discrete Ricci flow to calculate the Riemannian metric with a specified Gaussian curvature on the surface.

In differential geometry, the two principal curvatures of surface at a given point are the eigenvalues of the shape, which measure how the surface bends by different amounts in different directions at that point. The surface could be represented by the implicit function theorem as the graph of a function. Various global shape features can be derived by integrating local curvatures. When transfer the feature onto a 3D surface by directly mapping the feature image, the results are distorted obviously. It causes the difference between the geometric metrics of 3D surface and feature image. The total curvature of surface σ is determined by the topology of the surface. Given a mesh, we fix the connectivity and consider all possible metrics and the corresponding curvatures.

3.2 *Surface Parameterization Using Ricci Flow*

Ricci flow technique provides a mathematical tool to compute the desired metric that satisfies prescribed curvatures, induces zero Gaussian curvature when we determine a flat metric. In the discrete case, we embed the vertices of a mesh surface onto the plane using Euclidean Ricci flow method, proposed by [11], which can handle a surface with arbitrary topology and genus, as well this algorithm is convergence and stability, since the curvatures of singularities can be controlled completely. It has been proved that the surface parameterization can be calculated with a uniform metric. By embedding the vertices onto a plane with metric, we can obtain a conformal parameterization of a surface.

In this paper, we take over Jin et al. work [12] and use the discrete Euclidean Ricci flow to determine a flat metric that induces zero Gaussian curvature for 3D surface. According to this method, the surfaces with arbitrary topology and genus are possible to be handled, regardless of close or not. Moreover, since the states of singularities are completely configured, the surface boundaries can be stretched into any shapes. Here we support the metric of boundaries is constant geodesic curvature and interior is Gaussian curvature. If the Gaussian curvature equals zero, the mapping surface is Euclidean plane. Specially, the Euclidean Ricci flow for discrete surface is simply

$$\frac{d\mu_i}{dt} = \bar{G}_i - G_i, \tag{3}$$

which is the negative gradient flow of the function called Ricci energy

$$F(u) = -\int_0^u \sum_i \left(\bar{G}_i - G_i\right) du_i, \tag{4}$$

where $u = (u_1, u_2, \ldots, u_n)$. Its invert problem can be minimized by Newton's method. Jin *etal*. skillfully approximate the conformal metrics by conformal circle packing metrics. They map the surfaces in two phases: initialize a circle packing metric; and embed the mesh with the computed metric [10]. In parameter domain induced by a uniform flat metric, camouflage texture is mapped to the corresponding positions on surface shapes.

3.3 Parameterization Process

Circle Packing Method for conformal mapping was introduced by Thurston [8]. The first variational principle and different variational principles were proposed consequently, two of which are the foundation of computer implements.

Conformal metrics are approximated by conformal circle packing metrics. We associate each vertex v_i with a circle whose radius is γ_i (see Fig. 3). Let the set of radii $\Gamma = \{\gamma_i\}$ and the edge weights $\Theta = \theta_{ij}$ determine the metric of the mesh. (Γ, Θ) is called a circle packing metric of the mesh. Conformal maps transform infinitesimal circles to infinitesimal circles and preserve the intersection angles among the circles. Therefore, discrete conformal mappings modify the radii but preserve edge weights. Two circle packing metrics for the same mesh (Γ_1, Θ_1) and (Γ_2, Θ_2) are conformal to each other if and only if $\Theta_1 = \Theta_2$.

In order to parameterize a surface onto the plane, we firstly compute a circle packing metric (Γ, Θ), which approximates the original induced Euclidean metric on the surface as close as possible. Based on the equation below, we use a constrained optimization method to compute (Γ, Θ) by minimizing

$$E(\Gamma, \Theta) = \sum_{e_{i,j} \in E} l_{ij}^2 - \gamma_i^2 - \gamma_j^2 - 2\gamma_i\gamma_j \cos\phi_{ij}, \phi_{ij} \in \left[0, \frac{\pi}{2}\right], \tag{5}$$

where l_{ij} is the given edge length of the mesh. We use Newton's method to minimize this nonlinear energy [12]. If the initial triangulation does not contain

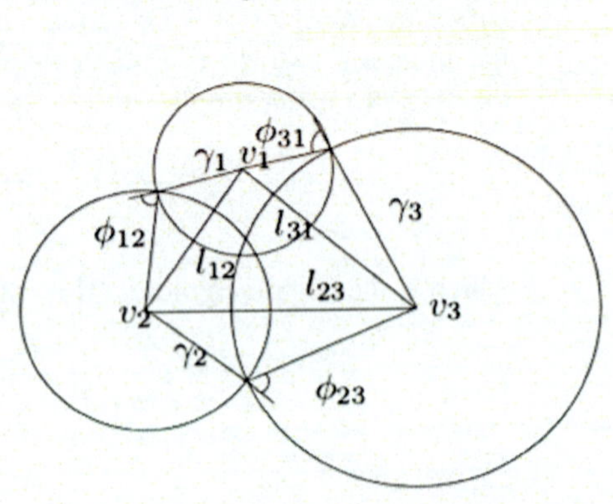

Fig. 3 A unit of CirclePacking on triangular mesh

many highly skewed triangles, we can obtain a circle packing metric close to the original Euclidean metric. Otherwise, we refine skewed triangles to improve the quality of the triangulation.

4 Experiments

4.1 Feature Synthesis Results

Losses are calculated by the following layers of the VGG-19 network: texture relu4_2 and feature relu1_1, relu2_1, relu3_1, relu4_1, relu5_1. We use L-BFGS energy to minimize the function. The stop arguments are the optimization has converged if the loss does not exceed 0.01% change during 50 iterations. For most stylizations, this is enough for about 1000 iterations, and in a few cases, more iterations will be needed, depending on the complexity of the feature image. If a convergence threshold of 0.1% is used, the number of iterations and run-time is halved, and we find that the effect is still acceptable. Two cases of feature synthesis results are shown in Fig. 4. Top row is in the case of woodland and bottom row is the synthesis result of rocky images.

4.2 3D Mapping Results

In the case of static target curvature , the Ricci flow algorithm can be applied directly. When the Euler number is positive, the static target curvature setting is similar to conventional parameterizations, where the parameter domain is a sphere or a polygon. Similar to the torus shape, all singularities need to be on the cut graph in order to obtain a valid parameterization. If a surface σ is with a non-zero Euler number, according to Gauss–Bonnet theory, some singularities should be contained at either interior or boundary vertices, where the curvatures are not zero. Figure 5 demonstrates two cases of camouflage mapping model results.

4.3 Camouflage Ability Results

In terms of the criteria of assessment, we flowed the inspire of [3] methods to collect the dependent factors, which included hit rates, detection time, the number of fixations, blinks, first cascade amplitude to interest objects (IOs), fixation duration, and fixation number in IOs. The definitions of these concepts are listed below. Eye fixation was calibrated manually at the beginning of experiment using fixation points presented randomly at different locations on the monitor. The accuracy of the

Fig. 4 A unit of CirclePacking on triangular mesh

Fig. 5 Two cases of feature mapping results

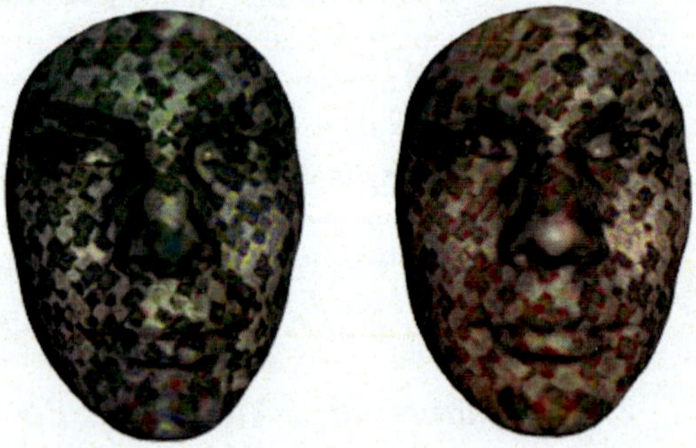

Fig. 6 Illumination of camouflage ability testing

Table 1 Testing results of camouflage ability

Background Hit rate difficulty			1st cascade amplitude to AI		
			Mean	Std. deviation	Confidence
Rocky	76	92	1.84	3.32	2.0±0.5
Woodland	165	41	3.21	6.28	6.1±0.6

calibration was checked between each trial, and the participants were asked to fixate on a fixation dot at the center of display to perform a correction for drifts and slips of the eye-tracker. The experiment had no time limit. If the participant could not find the target, he could press a key on the keyboard to quit the search. Participants were told that every task had a target, and that they were to find the camouflaged human shape and click the computer mouse to confirm it as quickly and accurately as possible. One experiment screen is shown in Fig. 6.

We tested the camouflage patterns of "87 Style" and "18 Style" which both are used in China Army. The detectability and discriminability are compared with our proposed pattern. The target search performance in camouflage environments was analyzed. The case of the grassy background is discussed first here. From the results of hit rate on the grassy background, the proposal camouflage pattern appeared to have good performance, and the other patterns had no significant differences, according to post hoc analysis. Table 1 presents the statistics between a hit and a miss for the first saccade amplitude to interest area. On both backgrounds, the first saccade amplitude could declare a hit very accurately, since the 95% confidence intervals did not overlap. For example, on the rocky background, the mean first saccade amplitude to interest area for a miss was 1.8432, with a 95% confidence interval between 0.87 and 1.17 (Table 2).

Table 2 Compare results of camouflage ability

Camouflage	Hit rate	Detection time	Fixations number	Blinks number	1st saccade Amp	Fixation duration	Fixations number
87 style	76	36014	104	4.10	4.86	2354	1.17
18 style	96	7854	25	0.47	7.32	1845	0.87
Proposal	96	4759	11	0.21	11.02	1652	1.04

5 Conclusion

By comparing several different camouflage design patterns against two different backgrounds, performance-based camouflage effectiveness data were collected. This study proposed a novel camouflage synthesis method, which significantly enhance the sensitivity of the comparisons between the different designs, according to the performance-based measures such as the hit rate and detection time. In particular, we extent the features map to 3D surfaces. From the analysis of the proposed camouflage patterns, it was found that detectability may play a more important role than discriminability. By means of artificial intelligent technology, the synthesis of camouflage pattern becomes flexible and efficient. Moreover, the adaptation process of camouflage pattern is real time, which makes the proposal is able to applied for adapted camouflage cover system.

Acknowledgments The support for this research is provided in part by the Tianjin Nature Science Foundation Project (18JCTPJC62400).

References

1. D. Lake, *The Army and Technology* (Palgrave Macmillan, New York, 2019)
2. J.C. Kilian, L.B. Hepfinger, Computer-based evaluation of camouflage. *Aerospace Sensing*, vol. 1687, pp. 106–114 (Proc. SPIE, Orlando, 1992). https://doi.org/10.1117/12.137864
3. M. Hogervorst, A. Toet, P.A.M. Jacobs, Design and evaluation of (urban) camouflage. *Infrared Imaging Systems: Design, Analysis, Modeling, and Testing XXI*, vol. 7662, p. 766205 (Proc. SPIE, Orlando, 2010). https://doi.org/10.1117/12.850423
4. N.E. Scott-Samuel, R. Baddeley, C.E. Palmer, I.C. Cuthill, Dazzle camouflage affects speed perception. PLoS ONE **6**(6), e20233 (2011). https://doi.org/10.1371/journal.pone.0020233
5. C. Lin, C. Chang, Y. Lee, Evaluating camouflage design using eye movement data. Appl. Ergon. **45**, 714–723 (2013). https://doi.org/10.1016/j.apergo.2013.09.012
6. X. Feng, S. Xu, Y.T. Luo, W. Jia, Design of digital camouflage by recursive overlapping of pattern templates. Neurocomputing **172**, 262–270 (2015). https://doi.org/10.1016/j.neucom.2014.12.108
7. X. Xue, F. Wu, J. Wang, Y. Hu, Camouflage texture design based on its camouflage performance evaluation. Neurocomputing **274**, S0925231217306732 (2018). https://doi.org/10.1016/j.neucom.2016.07.081
8. C. Li, M. Wand, Combining Markov random fields and convolutional neural networks for image synthesis. CoRR **abs/1601.04589** (2016). https://doi.org/10.1016/j.apergo.2013.09.012

9. L.A. Gatys, A.S. Ecker, M. Bethge, A neural algorithm of artistic style. CoRR **abs/1508.06576** (2015). https://doi.org/10.1117/12.850423
10. P. O'Donovan, A. Hertzmann, AniPaint: interactive painterly animation from video. IEEE Trans. Vis. Comput. Graph. **18**(3), 475–487 (2012)
11. B. Chow, F. Luo, Combinatorial ricci flows on surfaces. J. Differ. Geom. **63**(1), 97–129 (2003). https://doi.org/10.4310/jdg/1080835659
12. M. Jin, J. Kim, F. Luo, S. Lee, X. Gu, Conformal surface parameterization using euclidean ricci flow. Citeseerx (2012). https://doi.org/10.1.1.85.7199
13. H. Lu, Y. Li, S. Mu, D. Wang, H. Kim, S. Serikawa, Motor anomaly detection for unmanned aerial vehicles using reinforcement learning. IEEE Internet Things J. **5**(4), 2315–2322 (2018). https://doi.org/10.1109/JIOT.2017.2737479
14. H. Lu, Y. Li, T. Uemura, H. Kim, S. Serikawa, Low illumination underwater light field images reconstruction using deep convolutional neural networks. Futur. Gener. Comput. Syst. **82**, 142–148 (2018). https://doi.org/10.1016/j.future.2018.01.001
15. H. Lu, D. Wang, Y. Li, H. Kim, S. Serikawa, I. Humar, CONet: A cognitive ocean network. IEEE Wirel. Commun. **26**(3), 90–96 (2019)

Index

Y. Li, H. Lu (eds.), *3rd EAI International Conference on Robotic Sensor Networks*, EAI/Springer Innovations in Communication and Computing,
https://doi.org/10.1007/978-3-030-46032-7